Second Edition

BATCH DISTILLATION

Simulation, Optimal Design, and Control

Urmila Diwekar

CRC Press
Taylor & Francis Group
Boca Raton   London   New York

CRC Press is an imprint of the
Taylor & Francis Group, an **informa** business

CRC Press
Taylor & Francis Group
6000 Broken Sound Parkway NW, Suite 300
Boca Raton, FL 33487-2742

First issued in paperback 2017

© 2012 by Taylor & Francis Group, LLC
CRC Press is an imprint of Taylor & Francis Group, an Informa business

No claim to original U.S. Government works

Version Date: 20111109

ISBN 13: 978-1-4398-6122-6 (hbk)
ISBN 13: 978-1-138-07317-3 (pbk)

Visit the Taylor & Francis Web site at
http://www.taylorandfrancis.com

and the CRC Press Web site at
http://www.crcpress.com

To my parents, who taught me how to face challenges.

Contents

List of Figures

List of Tables

PREFACE TO SECOND EDITION

Last decade, a number of areas of batch distillation received considerable attention from researchers. Therefore, in this edition, I have added special sections on complex column configurations and azeotropic, extractive, and reactive distillation. A separate chapter on uncertainties in batch distillation is also added to this new edition. This deals with important and recent developments dealing with various kinds of uncertainties in batch distillation. The total number of chapters increased from seven to ten, as the chapter on complex columns and complex systems was divided into two separate chapters. Chapter 10 is devoted to various software packages for batch distillation simulation, design, optimization, and control. I tried to cover most of the literature published so far. As time proceeds I am sure more research will be added to this fascinating area of batch distillation.

Urmila M. Diwekar
Clarendon Hills, Illinois

PREFACE TO FIRST EDITION

The recent increase in the production of high-value-added, low-volume specialty chemicals and biochemicals has generated a renewed interest in batch processing technologies. Batch distillation is an important unit operation in the batch processing industry and is widely used. The flexibility of batch distillation combined with the inherent unsteady nature of the process poses challenging design and operation problems. In view of the practical importance and in response to industrial needs for chemical engineers with a strong background in batch processing, more and more educational institutions are redesigning their curricula to include courses devoted to the subject. From both academic and industrial standpoints, therefore, a book dedicated to the subject of batch distillation is of much significance, which is the motivation for the present undertaking. Because the existing books in chemical engineering pertain mainly to continuous processes, with the subject of batch distillation being relegated to only a small section, the book fills an important void in chemical engineering literature.

The book features the following:

(a) It introduces the various operating modes in detail.

(b) It examines the challenges involved in a rigorous modeling of batch distillation column dynamics.

(c) It provides a hierarchy of models of varying complexity and rigor.

(d) It presents approaches to optimal design of batch distillation columns and highlights the differences vis-a-vis continuous columns.

(e) It describes optimal control problems in batch distillation.

(f) It discusses analysis and synthesis of columns with complex thermodynamics, and complex, unconventional column configurations.

In addition to presenting an overview of batch distillation design, operation, and control, the book provides a basic understanding of the operation of batch and semi-continuous units, design methods, and solution techniques for handling unit complexities, and batch processes in general.

The book is intended for a diverse audience ranging from undergraduate and graduate students of chemical engineering to researchers and academicians faced with batch distillation research problems, and practicing chemical engineers interested in tackling problems encountered in actual operations. It is the objective of the endeavor to provide a timely source and reference material for educational use in a chemical engineering curriculum. This book is primarily designed to serve as a textbook for a graduate course in batch distillation. However, the early chapters (Chapters 1 and 2) contain material that can be used in undergraduate chemical engineering courses as well. In the course of the preparation of the book, a preliminary draft of the manuscript was used in a batch design course taught at the Massachusetts Institute of Technology, during fall 1994.

Aside from its use in courses, the book will serve as a comprehensive reference book for chemical engineering professionals. This book will enable readers to effectively design, synthesize, and make operations decisions related to batch processes. The understanding of optimization and optimal control methods will provide the necessary tools for making best decisions in practice. To readers in academia and the industries alike, I hope that the ideas and pages that follow will give as much enjoyment and stimulate creative interest as they have for me right from the day I started working in this area.

Acknowledgments

> *And you receivers—and you are all receivers—assume no weight of gratitude, lest you lay a yoke upon yourself and upon him who gives.*
> *Rather rise together with the giver as on his gifts as on wings;*
> *For to be mindful of your debt is to doubt his generosity who has the free-hearted earth for mother, and God for father.*
>
> *– Khalil Gibran*

This book would not exist without the personal and professional support of "Pitchu" (Dr. Rangarajan Pitchumani, University of Connecticut). He was there for me in every challenge I faced, including this one, starting from my initial days in the United States. My heartfelt thanks to him for providing me a "home away from home" and for sharing my anxieties in crossing every hurdle during my stay in this country.

> *Your friends are your needs answered*
>
> *– Khalil Gibran*

Thanks to Dr. Narasimhan Devanathan (Amoco Corporation) whose suggestion of the idea of writing this book set me started on the project. Thanks also go to the many who shared my pleasures and depressions while working on this book.

I am very appreciative of the efforts of Dr. Edward Rosen, CACHE Corporation; Dr. Rafiqul Gani, Technical University of Denmark; Dr. Greg McRae, MIT; Dr. Claudia Schmidt and Dr. Vasant Shah, Simulation Sciences, Inc.; Dr. Lorenz Biegler, Carnegie-Mellon University; and Ravi Gudi, University of Alberta, Canada, for their many valuable suggestions. I am grateful to

Carnegie-Mellon graduate students Ramesh Iyer, Boyd Safrit, and Prosenjit Chaudhuri for the careful review of the initial drafts of the manuscript and for their comments. Thanks again to Pitchu for meticulously correcting all the different versions of the manuscript, checking technical derivations for "making physical sense," and helping me in every possible way to complete this book.

Of course, I am especially thankful to my dearest and nearest family members Anjali Diwekar, Ashish Joshi, and Sanjay Joag for their support and patience. Their enthusiasm and admiration of even very small advancements in my research and other endeavors made all the efforts worthwhile.

To my parents, for their untiring love and caring, and for all the sacrifices behind their greatest gift to me—education, I dedicate this work.

Urmila M. Diwekar
Pittsburgh, Pennsylvania
May 1995

NOTATIONS

B amount remaining in the reboiler [mole]

B_t amount remaining in the reboiler at time t [mole]

c_1 amortized incremental investment cost [\$/m^2/plate/year]

c_2 amortized incremental tubular equipment cost [\$/m^2/yr]

c_3 cost of the steam and coolant to vaporize or condense, respectively, 1 kg of distillate [\$/kg]

C_0 cost of feed [\$/mole]

C_1 constant in the Hengestebeck–Geddes equation for conventional batch column

C_B constant in the Hengestebeck–Geddes equation for stripper or stripping section

C_j conversion fraction of reaction on plate j

C_T constant in the Hengestebeck–Geddes equation for rectifying section

CF capacity factor [mole/hr]

D total distillate [mole]

D_t total distillate at time t [mole]

Da *Damköhler* number

F total feed [mole]

G_a allowable vapor velocity [kg/hr/m^2]

G_b vapor handling capacity of the equipment [kg/hr/m^2]

H_c molar compartmental holdup [mole]

H_B molar reboiler holdup [mole]

H_D molar condenser holdup [mole]

H_j molar holdup on plate j [mole]

H_m middle vessel molar holdup [mole]

H_{mix} enthalpy of mixing [kJ]

HHK heavier than heavy key component

HK heavy key component

HRs hrs of operation per year [hr/yr]

I_B enthalpy of liquid in the reboiler [J/mole]

I_B enthalpy of liquid in the condenser [J/mole]

I_j enthalpy of the liquid stream leaving plate j except for the collocation model in Chapter 4 [J/mole]

I_{j-1} enthalpy of the liquid stream entering plate j except for the collocation model in Chapter 4 [J/mole]

I_m enthalpy of liquid in the middle vessel [J/mole]

J_B enthalpy of vapor in the reboiler [J/mole]

J_j enthalpy of the vapor stream leaving plate j except for the collocation model in Chapter 4 [J/mole]

J_{j+1} enthalpy of the vapor stream entering plate j except for the collocation model in Chapter 4 [J/mole]

$K_1,\ K_2,\ K_3$ cost coefficients

K_j kinetic parameter given by Arrehenius equation on plate j

L_j liquid stream leaving plate j except for the collocation model in Chapter 4 [mole/hr]

L_{j-1} liquid stream entering plate j except for the collocation model in Chapter 4 [mole/hr]

LK light key component

LLK lighter than light key component

n number of component

N number of theoretical plates

N_B number of theoretical plates in stripper or stripping section

N_{min} minimum number of plates

Nb_{min} minimum number of plates in stripper or stripping section (bottom section)

$ncol$ number of collocation points

Nt_{min} minimum number of plates in rectifying or top section

N_{minf} minimum number of plates at the terminal condition

N_T plate corresponding to the middle vessel

$ncol$ number of collocation points

P pressure [atm]

P_j pressure on plate j except for the collocation model in Chapter 4 [atm]

P_r sales value of the product [\$/mole]

q feed thermal condition

Q_B reboiler heat duty [J/hr]

Q_D condenser heat duty [J/hr]

Q_m middle vessel heat duty [J/hr]

Q_R total reboiler heat duty [kJ]

Q_s total still heat duty in the stripper [kJ]

$r_{k,j}$ rate of reaction of component k on plate j

R reflux ratio

r_0 reference rate of reaction

R_{max} maximum initial reflux ratio for the variable reflux mode

R_{min} minimum reflux ratio

R_{MIN} minimum feasible reflux ratio for the constant reflux and optimal reflux operations

$R_{min\ g}$ minimum reflux ratio given by the Gilliland correlation

$R_{min\ u}$ minimum reflux ratio given by the Underwood equations

R_t — reflux ratio at time t

Rb — reboil ratio

Rb_{min} — minimum reboil ratio

$Rb_{min\ g}$ — minimum reboil ratio given by the modified Gilliland correlation

$Rb_{min\ u}$ — minimum reboil ratio given by the Underwood equations

RR — dimensionless rite of reaction

S — pole height in Chapter 2; amount remaining in the still for stripper in Chapter 6

T — total batch time [hr]

T_{eq} — equilibration time [hr]

t_s — setup time [hr]

TE — temperature [degrees K]

TE_j — temperature on plate j except for the collocation model in Chapter 4 [degrees K]

x_B — liquid composition of component 1 in a binary mixture [mole fraction]

$x_b^{(i)}$ — compartmental output liquid composition of component i [mole fraction]

$x_B^{(i)}$ — liquid composition of component i in the reboiler [mole fraction]

$x_{B0}^{(i)}$ — initial reboiler composition of component i [mole fraction]

$x_{B\infty}^{(1)}$ — final reboiler composition of component 1 [mole fraction]

x_D — distillate composition of component 1 in a binary mixture [mole fraction]

x_D^* — specified average distillate composition of the key component [mole fraction]

$x_D^{(i)}$ — distillate composition of component i [mole fraction]

$x_{D0}^{(i)}$ — initial distillate composition of component i [mole fraction]

x_{Dav} — average distillate composition of the key component [mole fraction]

x_F — feed composition of component 1 in a binary mixture [mole fraction]

$x_F^{(i)}$ — feed composition of component i [mole fraction]

x_j — liquid composition of component 1 in a binary mixture leaving plate j except for the collocation model in Chapter 4 [mole fraction]

x_{j-1} — liquid composition of component 1 in a binary mixture entering plate j except for the collocation model in Chapter 4 [mole fraction]

$x_j^{(i)}$ — liquid composition of component i leaving plate j except for the collocation model in Chapter 4 [mole fraction]

$x_{j-1}^{(i)}$ liquid composition of component i entering plate j except for the collocation model in Chapter 4 [mole fraction]

$x_m^{(i)}$ middle vessel liquid composition of component i [mole fraction]

$x_s^{(i)}$ composition of component i on the sensitive plate in Chapter 4; still composition of stripper in Chapter 6 onward [mole fraction]

$x_{t-1}^{(i)}$ compartmental input liquid composition of component i [mole fraction]

V vapor boilup rate [mole/hr]

V_B vapor stream leaving the reboiler [mole/hr]

V_j vapor stream leaving plate j except for the collocation model in Chapter 4 [mole/hr]

V_{j+1} vapor stream entering plate j except for the collocation model in Chapter 4 [mole/hr]

V_T vapor boilup rate at the top section of the middle vessel column [mole/hr]

Vol volumetric flowrate entering a plate

y_B vapor composition of component 1 in a binary mixture which is in equilibrium in the reboiler [mole fraction]

$y_B^{(i)}$ vapor composition of component i which is in equilibrium in the reboiler [mole fraction]

$y_{b+1}^{(i)}$ compartmental input vapor composition of component i [mole fraction]

y_j vapor composition of component 1 in a binary mixture leaving plate j except for the collocation model in Chapter 4 [mole fraction]

y_{j+1} vapor composition of component 1 in a binary mixture entering plate j except for the collocation model in Chapter 4 [mole fraction]

$y_j^{(i)}$ vapor composition of component i leaving plate j except for the collocation model in Chapter 4 [mole fraction]

$y_{j+1}^{(i)}$ vapor composition of component i entering plate j except for the collocation model in Chapter 4 [mole fraction]

$y_t^{(i)}$ compartmental output vapor composition of component i [mole fraction]

Greek symbols

α relative volatility of component 1 in a binary mixture

α_i relative volatility of component i

γ_i stoichiometric coefficient of reactant i (−ve for reactant, +ve for product)

γ_T sum of total stochiometric coefficient

ϵ_{kj} sum of molar turnover flowrate for component k on plate j

ϵ_{kT} total molar turnover flowrate for component k

λ latent heat of vaporization [kJ/mole]
     Lagrange multiplier

λ_i Lagrange multiplier for constraint i
     eigenvalue

ρ molar density [mole/vol]

ϕ Underwood constant

$\triangle n_B$ total change in moles due to reaction in reboiler

$\triangle n_D$ total change in moles due to reaction in condenser

$\triangle n_j$ total change in moles due to reaction on plate j

1

INTRODUCTION

CONTENTS

Distillation is the oldest separation process and the most widely used unit operation in industry. Distillation began as a simple still in a laboratory. Figure 1.1 shows a schematic illustration of such a still initially filled with a feed mixture which evaporates and leaves the liquid. The vapor, which is richer in the more volatile component, is collected in the condenser at the top. The simple distillation still is an example of a batch operation. The concept of reflux and the use of accessories such as packings or plates to increase the mass transfer convert this simple still into a batch distillation column.

The shift from batch to continuous operation began in petrochemical industries and slowly continuous distillation came to dominate all sections of the large-scale industry. Small-scale, specialty chemical, and pharmaceutical industries, however, still used batch distillation columns and do so to this day. Moreover, the recent increase in the production of high-value-added, low-volume specialty chemicals and biochemicals has generated a renewed interest in batch distillation. The flexibility offered by batch distillation gives it an advantage over continuous distillation. At the same time, this flexibility and the inherent unsteady state nature of batch distillation pose challenging design and operational problems. This chapter introduces batch distillation, presents an analysis of a simple batch distillation process, and outlines the advantages of this unit operation.

The analysis of simple distillation presented in 1902 by Lord Rayleigh (Rayleigh, 1902) marks the earliest theoretical work on batch distillation. The following subsection presents this early analysis of the process.

1.1   Simple Distillation

Simple distillation, also called differential distillation, is the most elementary example of batch distillation. As shown in Figure 1.1, in this process the vapor is removed from the still during each time interval and is condensed in the condenser. The vapor is richer in the more volatile component than the liquid remaining in the still. Over time, the liquid remaining in the still becomes weaker in the concentration of the more volatile component, while the distillate collected in the condenser gets progressively enriched in the more volatile component. The analysis of this process for a binary system, proposed by Lord Rayleigh, is given below.

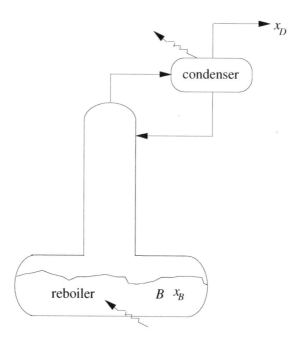

FIGURE 1.1
Schematic of a Simple Distillation Operation

Let F (moles) be the initial binary feed to the still and x_F (mole fraction) be the composition of component A in the feed. Let B be the number of moles of material remaining in the still, x_B the mole fraction of component A in the still, and x_D the mole fraction of component A in the vapor dB produced during an infinitesimal time interval dt.

The differential material balance for component A can then be written as:

$$
\begin{aligned}
x_D dB &= d(B x_B) = B dx_B + x_B dB \\
\int_F^B \frac{dB}{B} &= \int_{x_F}^{x_B} \frac{dx_B}{x_D - x_B} \\
\ln\left(\frac{B}{F}\right) &= \int_{x_F}^{x_B} \frac{dx_B}{x_D - x_B}
\end{aligned}
\tag{1.1}
$$

In the distillation process, it is assumed that the vapor formed within a short period is in thermodynamic equilibrium with the liquid. Hence, the vapor composition x_D is related to the liquid composition x_B by an equilibrium relation of the functional form $x_D = f(x_B)$. The exact relationship for a particular mixture may be obtained from a thermodynamic analysis and is also dependent upon temperature and pressure. Figure 1.2 shows an example equilibrium curve for a system consisting of CS_2 and CCl_4 at 1 atmosphere pressure. Since the aim of this book is to explain the concepts behind batch distillation operations, simple equilibrium relations based on ideal systems will be used in most parts of the book, except for certain chapters and exercises.

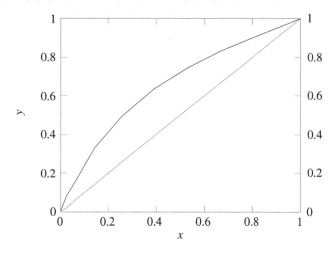

FIGURE 1.2
Equilibrium Curve for the CS_2 and CCl_4 Mixture at 1 Atmosphere Pressure

If the equilibrium relationship is a straight line (which is the case in ideal dilute mixtures) then the Rayleigh equation, Equation 1.1, reduces to the following form.

$$\frac{x_D - x_B}{x_{D0} - x_F} = \left(\frac{B}{F}\right)^{m-1} \qquad (1.2)$$

where m is the slope of the equilibrium line and x_{D0} is the initial composition of the more volatile component in the distillate.

Similarly, for a system following the ideal behavior given by Raoult's law, the equilibrium relationship between the vapor composition y (x_D) and liquid composition x (x_B) of the more volatile component in a binary mixture can be approximated using the concept of constant relative volatility (α), and is given by:

$$y = \frac{\alpha x}{(\alpha - 1)x + 1} \qquad (1.3)$$

Substitution of the above equation in Equation 1.1 results in:

$$\ln\left(\frac{B}{F}\right) = \frac{1}{\alpha - 1}\ln\left[\frac{x_B(1 - x_F)}{x_F(1 - x_B)}\right] + \ln\left[\frac{1 - x_F}{1 - x_B}\right] \qquad (1.4)$$

Example 1.1: A liquid mixture containing 50 mole percent benzene and 50 mole percent toluene is charged in the simple distillation still. How much charge must be boiled away to leave a mixture containing 80 mole percent toluene? Assume that the relative volatility for the system, α_{b-t}, is 2.4.

Solution: Since the relative volatility of the mixture is a constant, Equation 1.4 can be used. Equation 1.4 is derived in terms of the more volatile component. The following calculations are carried out using benzene compositions.

$$\ln\left(\frac{B}{F}\right) = \frac{1}{1.4}\ln\left[\frac{(0.2)(0.5)}{(0.5)(0.8)}\right] + \ln\left[\frac{0.5}{0.8}\right] = -2.5376$$

$$\frac{B}{F} = 0.0791$$

Therefore, $100 \times (1 - \frac{B}{F})$ equal to 92.1 percent of the initial liquid must be boiled away.

Although simple distillation historically represents the start of the distillation process, a complete separation using this process is impossible unless the relative volatility of the mixture is infinite. Therefore, the application of these stills is restricted to laboratory scale distillation, where high purities are not required or when the mixture is easily separable.

One can look at simple distillation as consisting of one equilibrium stage where a liquid and a vapor are in contact with one another and the transfer takes place between the two phases, as shown in Figure 1.3a. If N such stages are stacked one above the other, as shown in Figure 1.3b, and are allowed to have successive vaporization and condensation, it results in a substantially richer vapor and weaker liquid, in terms of the more volatile component, in the condenser and the reboiler, respectively. This multistage arrangement, shown in Figure 1.3b, is representative of a distillation column, where the vapor from the reboiler rises to the top and the liquid from the condenser is refluxed downwards. The contact between the liquid and the vapor phase is established through accessories such as packings or plates. However, it is easier to express the operation of the column in terms of thermodynamic equilibrium stages which represent the theoretical number of plates in the column. Therefore, we will restrict our discussions to these plate columns.

Figures 1.4 and 1.5 show a continuous and a batch operation of this multistage process. A distillation column in general consists of a cylindrical structure divided into sections by a series of perforated plates or trays that permit the upward flow of vapor. The liquid reflux flows across each plate over a weir and down a downcomer to the plate below. This liquid is the refluxed portion of the vapor, which has reached the top and condensed in the condenser. The vapor and liquid come in contact on each plate or stage where the mass

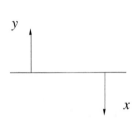

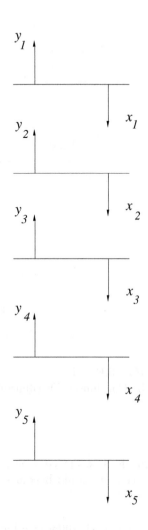

FIGURE 1.3
Equilibrium Stage Processes: (LHS) Single Stage Process, (RHS) Multistage
Process

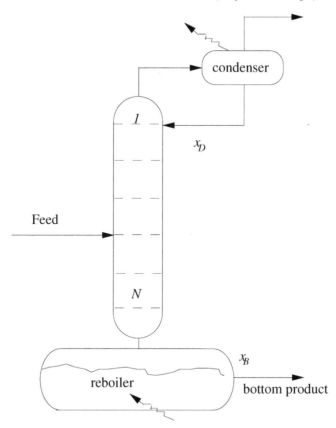

FIGURE 1.4
The Continuous Distillation Column

transfer takes place. Consequently, the rising vapor becomes richer, while the descending liquid becomes correspondingly weaker in the more volatile component.

The basic difference between a batch column and a continuous column is that in continuous distillation the feed is continuously entering the column, while in batch distillation the reboiler is normally fed at the beginning of the operation. Also, while the top products are removed continuously in both batch and continuous operations, there is no bottom product in batch distillation. Since in a continuous operation the total product flow rate equals that of incoming feed or feeds, the process reaches steady state. In batch distillation, on the other hand, the reboiler gets depleted over time, so the process is unsteady.

1.2   Why Batch Distillation?

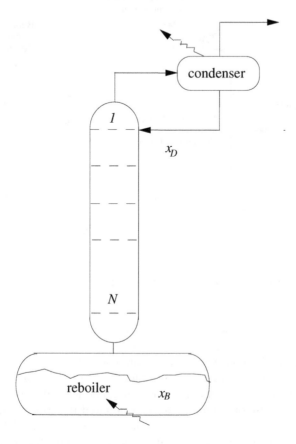

FIGURE 1.5
The Batch Distillation Column

Batch distillation is an important unit operation frequently used for small-scale production. Batch distillation is preferable to continuous distillation when small quantities of high technology/high-value-added chemicals and bio-chemicals need to be separated. The most outstanding feature of batch distillation is its flexibility. This flexibility allows one to deal with uncertainties in feed stock or product specification. Also, one can handle several mixtures just by switching the column's operating conditions, a simple procedure.

Batch distillation requires the least amount of capital for separating relatively pure components. Continuous distillation generally requires a separate column for each component. Figure 1.6 shows the two-column separation sequence for separating benzene-toluene-xylene (Nishimura and Hiraizumi,

1971). The same separation could be obtained with one batch distillation column, shown in Figure 1.7, with different product cuts or fractions.

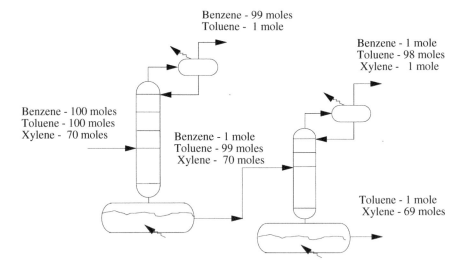

FIGURE 1.6
Sequence of Continuous Columns for Separation of Benzene–Toluene–Xylene System

   Although it is proposed that to separate n pure components, one may need $n-1$ continuous distillation columns (Nishimura and Hiraizumi, 1971), sometimes one needs at least $n-1$ continuous columns when the purity requirements dictate columns with an infinite number of plates or an infinite reflux ratio in separating the n component mixture. The feasibility of a given separation using a continuous column can be evaluated in terms of the minimum number of plates N_{min} and the minimum reflux ratio R_{min}. The minimum number of plates is defined as the number of equilibrium stages required for a given separation under the total reflux condition (infinite reflux ratio, where the reflux ratio is defined as the ratio of the liquid reflux rate at the top to the product rate in the condenser, $R = L/\frac{dD}{dt}$). Fenske (1932) derived an equation for calculating the minimum number of plates for a given separation using a continuous column when the mixture has a constant relative volatility. Fenske's equation for continuous distillation for a binary system can be given as follows:

$$N_{min} = \frac{\ln\left[\frac{x_D}{1-x_D}\frac{1-x_B}{x_B}\right]}{\ln[\alpha]} \qquad (1.5)$$

where x_D and x_B are, respectively, the distillate composition and the bottom composition of the more volatile component. However, the flexibility and the transient nature of batch distillation allow for the same separation using a

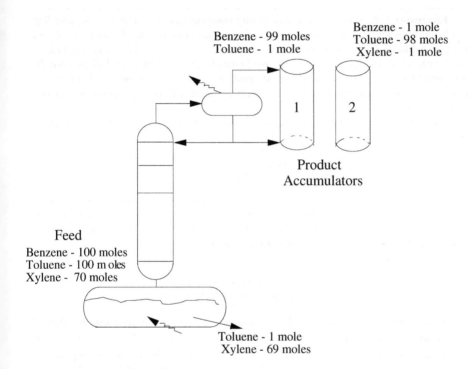

FIGURE 1.7
A Multifraction Batch Distillation Column for Separation of Benzene–Toluene–Xylene System

single column. For batch distillation Fenske's equation reduces to (for details, please refer to Chapter 4)

$$N_{min} = \frac{\ln\left[\frac{x_D}{1-x_D}\frac{1-x_F}{x_F}\right]}{\ln[\alpha]} \tag{1.6}$$

The following simple example illustrates this concept. This example is derived from an industrial batch distillation column currently operational at International Flavors and Fragrance (IFF) in New Jersey.

Example 1.2: A binary feed containing components A and B is to be distilled to obtain a relatively pure mixture containing component B. The distillation column that is available for the operation has six theoretical plates. The relative volatility of A with respect to B is 2.6, and the feed contains 30 percent A and 70 percent B. The product purity requirements are 97 percent B and 95 percent A. If the choice is between operating this column as a continuous or a batch operation, which one would you choose?

Solution: Using Fenske's equation, the minimum number of plates for the given separation using a continuous column is:

$$N_{min} = \frac{\ln\left[\frac{0.95}{1-0.95}\frac{1-0.03}{0.03}\right]}{\ln[2.6]} = 6.72$$

Since the minimum number of plates is more than the specified number of plates for the column operation, namely six, it is not possible to use a continuous column for this operation. The continuous operation is only feasible if one increases the number of plates or uses a sequence of continuous columns.

For batch distillation:

$$N_{min} = \frac{\ln\left[\frac{x_D}{1-x_D}\frac{1-x_F}{x_F}\right]}{\ln[\alpha]}$$

A preliminary analysis using the above relationship for the minimum number of plates for a given separation in a batch distillation results in $N_{min} = 3.97$ for a batch distillation column operating at a constant reflux. This makes the separation feasible using the given column with the number of plates equal to six. The feasibility considerations for batch distillation are described in Chapters 4 and 5 using a shortcut method.

The transient nature of batch distillation allows it to configure the column in a number of different ways, some of which are shown in Figure 1.8. The column in Figure 1.8a, as explained, is the conventional batch distillation column with the reboiler at the bottom and the condenser at the top, which essentially performs the rectifying operation. A single column can be used to separate several products using the multifraction operation of batch distillation presented in Figure 1.8b. Some cuts may be desired and others may be intermediate products. These intermediate fractions can be recycled to maximize profits and/or minimize waste. Figure 1.8c shows a periodic operation in which each charge consists of a fresh feed stock mixed with the recycled off-specification material from the previous charge. Configurations in Figure 1.8b and 1.8c can be implemented in an optimal campaign structure (Wajge and Reklaitis, 1998). Figure 1.8d represents a stripping column for separating the heavy product as the bottom product where the liquid feed is initially charged into the top. In 1994 Devidyan et al. presented a batch distillation column that has both stripping and rectifying sections embedded in it (Figure 1.8e), namely, the middle vessel column. Although this column has not been investigated completely, recent studies demonstrated that it provides added flexibility for the batch distillation operation. In 1997, Skogestad et al. described a new column configuration called a multivessel column (Figure 1.8f), which is similar to the MEBDS (multi-effect batch distillation system) of Hasebe et al.(1997), and showed that the column can obtain purer products at the end of a total reflux operation. These emerging designs play an important role in separation of complex systems like azeotropic, extractive, and reactive batch distillation systems. The batch rectifier configuration for such separations may be very restrictive and expensive. These are a few examples of the different kinds of batch distillation columns. Combined with different possible operating modes, the number of column configurations tends to be very high. However, in current practice the convention batch column (the rectifier) is commonly used. Therefore, I have devoted the first four chapters to this column. The next chapter describes the basic operating modes of this conventional batch distillation column.

References

Fenske M. R. (1932), Fractionation of straight run Pennsylvania gasoline, *Industrial and Engineering Chemistry*, **24**, 482.

Devidyan A. G., V. N. Kiva, G. A. Meski, and M. Morari (1994), Batch distillation in a column with a middle vessel, *Chemical Engineering Science*, **49**, 3033.

Hasebe S., M. Noda, I. Hashimoto (1997), Optimal operation policy for multi-effect batch distillation system, *Comp. Chem. Eng.*, **21**, s1221.

Kim K. and U. Diwekar (2001), New era in batch distillation: computer aided analysis, design, and control, invited paper, *Reviews in Chemical Engineering*, **17 (2)**, 111.

Nishimura H. and Y. Hiraizumi (1971), Optimal system pattern for multicomponent distillation, *International Chemical Engineering*, **11(1)**, 188.

Rayleigh, Lord (1902), On the distillation of binary mixtures, *Philosophical Magazine (vi)*, **4**, No. 23, 521.

Skogestad S., B. Wittgens, R. Litto, and E. Sørensen (1997), Multivessel Batch Distillation,*AIChE J.*, **43**, 971.

Wajge R. M. and G. V. Reklaitis (1998), An optimal campaign structure for multicomponent batch distillation with reversible reaction, *Ind. Eng. Chem. Res.*, **37**, 1910.

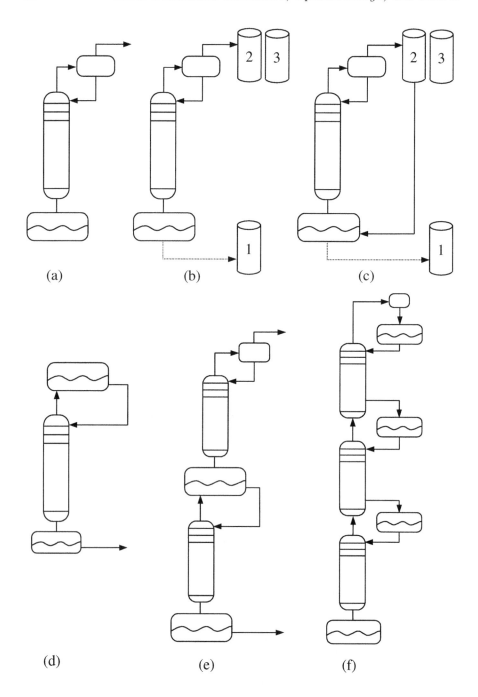

FIGURE 1.8
Examples of Ways to Configure the Column (Kim and Diwekar, 2001).

Exercises

1.1 Show that the Rayleigh equation reduces to Equation 1.2 for the linear equilibrium curve and to Equation 1.4 for the constant relative volatility model.

1.2 Modify the Rayleigh equation for a multicomponent mixture, assuming the constant relative volatility model given below for n components.

The equilibrium curve for the n component mixture is given by:

$$y_i = \frac{\alpha_i x_i}{\sum_{j=1}^{n} \alpha_j x_j}$$

1.3 A liquid mixture containing 60 mole percent benzene and 40 mole percent toluene is charged in a simple distillation still. Find the composition of the vapor accumulated in the condenser and the liquid remaining in the still when exactly half of the mixture is in the condenser. Assume the relative volatility α_{b-t} is 2.4.

1.4 An equimolar multicomponent mixture containing meta-, ortho-, and para-mono-nitro-toluene is to be distilled using differential distillation. Use the Rayleigh equation for a multicomponent mixture derived in Example 1.2 to find the resulting condenser and still charge compositions when the still has 40 percent of the initial charge left. Assume the relative volatilities α_{m-p} and α_{o-p} are 1.7 and 1.16, respectively.

1.5 The equilibrium data for a binary mixture of acetone and acetic acid is given below. If the acetone composition in the vapor at the end of the operation is 55 percent when the equimolar feed is distilled in the Rayleigh still, what percentage of the initial charge is in the condenser?

$x_{acetone}$: .050 .100 .200 .300 .400 .500 .600 .700 .800 .900
$y_{acetone}$: .162 .306 .557 .725 .840 .912 .947 .969 .984 .993

1.6 Define the minimum number of cuts required to separate a mixture of propane, butane, pentane, hexane, heptane, and octane into pure component products.

2

BASIC MODES OF OPERATION

CONTENTS

Chapter 1 introduced the batch distillation operation, described concepts behind the simplest batch distillation operation, and presented the advantages of using batch distillation columns. This chapter introduces and presents an analysis of the basic modes of batch distillation operation. First, a theoretical analysis is established, based on simplifying assumptions. Then, a graphical approach to batch distillation column design is developed and illustrated with simple examples of binary systems.

2.1 Theoretical Analysis

Figure 2.1 represents a schematic of a batch distillation column, where the total batch is fed to the reboiler at the bottom of the column. The column consists of perforated plates that permit the upward flow of vapor. The liquid from the top is refluxed back across each plate, and the vapor rising from the top plate passes to a condenser. In this chapter we will ignore the total liquid and vapor holdup on each plate and in the condenser. In Chapter 3, we will relax this assumption to include the holdup effects and will also discuss where it is necessary to use this zero holdup model. Under the simplification of negligible holdup, the overall material balance and the material balance for the more volatile component around the complete column, shown in Figure 2.1, result in the following equations.

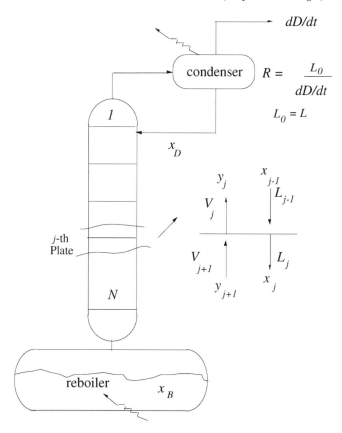

FIGURE 2.1
Schematic of a Batch Distillation Column

$$x_D dB = d(Bx_B) = B dx_B + x_B dB$$
$$\int_F^B \frac{dB}{B} = \int_{x_F}^{x_B} \frac{dx_B}{x_D - x_B}$$
$$\ln\left(\frac{B}{F}\right) = \int_{x_F}^{x_B} \frac{dx_B}{x_D - x_B} \tag{2.1}$$

In the above equation B is the amount remaining in the still, F is the initial feed, x_D is the instantaneous distillate composition, and x_B is the still composition of the more volatile component.

In Chapter 1, the Rayleigh equation was derived for a simple distillation still. It can be easily seen from the above analysis that the negligible holdup batch distillation column also follows the Rayleigh equation for simple distillation. The difference between the two processes, however, exists in the functional relationship between distillate composition x_D and bottom composition x_B.

Functional Relationship Between x_D and x_B

In the simple distillation analysis described in Chapter 1, the equilibrium relation curve represents the functional relationship between x_D and x_B. For the plate column considered here, the relationship requires a calculation of material and energy balance across each plate.

In Figure 2.1, the plates are numbered starting from the condenser, i.e., from the top of the column. For a control volume around the j-th plate there are four streams involved: liquid stream L_{j-1} and vapor stream V_{j+1} are entering and liquid stream L_j and vapor stream V_j are leaving the j-th plate. If the liquid composition and the vapor composition of the more volatile component are denoted by x and y, respectively, and the liquid and vapor enthalpies by I and J, respectively, the material and energy balances for the control volume can be written as follows:

Material balances:

$$L_{j-1} + V_{j+1} = L_j + V_j \tag{2.2}$$
$$L_{j-1}x_{j-1} + V_{j+1}y_{j+1} = L_jx_j + V_jy_j \tag{2.3}$$

Energy balance:

$$L_{j-1}I_{j-1} + V_{j+1}J_{j+1} = L_jI_j + V_jJ_j + \text{losses} + H_{mix} \tag{2.4}$$

where H_{mix} is the heat of mixing.

If the molar latent heat of vaporization is assumed to be constant ($J_{j+1} = J_j$), and the system is assumed to be ideal ($H_{mix} = 0$, losses $= 0$), then the energy balance equation (Equation 2.4) reduces to $L_{j-1} = L_j = L$ and $V_{j+1} = V_j = V$, which is the equimolar overflow assumption. The equimolar overflow assumption allows one to extend the popular McCabe and Thiele graphical method (McCabe and Thiele, 1925) for continuous distillation to batch distillation as shown below.

In the McCabe and Thiele method the material balance is considered around the condenser and the j-th plate. This leads to the following equations.

The overall material balance around the top section and the j-th plate:

$$V = L + dD/dt = dD/dt(R + 1) \tag{2.5}$$

where L represents the liquid refluxed from the condenser, R represents the reflux ratio defined as $R = \frac{L}{dD/dt}$, and dD/dt is the distillate rate.

The material balance for the more volatile component considering the envelope around the top section of the column and plate j in Figure 2.1 is

$$y_{j+1} = \frac{L}{V}x_j + \frac{dD/dt}{V}x_D \tag{2.6}$$

By substituting for V from Equation 2.5, Equation 2.6 can be expressed in terms of the reflux ratio R as follows.

$$y_{j+1} = \frac{R}{R+1}x_j + \frac{1}{R+1}x_D \qquad (2.7)$$

McCabe and Thiele proposed that since the above equation represents a straight line connecting y_{j+1} and x_j, it can be drawn on the same diagram as that of the equilibrium curve, shown in Figure 2.2. Recall that the equilibrium curve represents the relation between y_j and x_j, which in general may be expressed as:

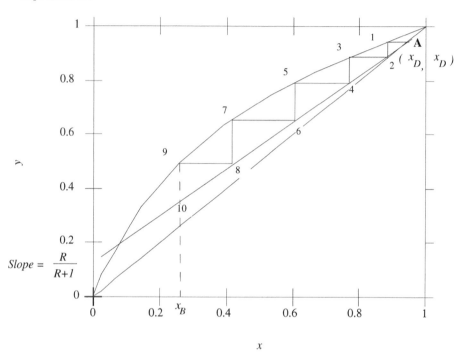

FIGURE 2.2
McCabe–Thiele Method for Plate-to-Plate Calculations

$$y_j = f(x_j) \qquad (2.8)$$

Thus, the line of Equation 2.7 will pass through points 2, 4, 6, etc., in Figure 2.2. In Equation 2.7, if $x_0 = x_D$ (the equation for the top plate of the column), then:

$$y_1 = \frac{R}{R+1}x_D + \frac{1}{R+1}x_D \qquad (2.9)$$

Therefore, Equation 2.7 represents a line through point $y_{j+1} = x_j = x_D$

with a slope $\frac{R}{R+1}$. Starting from this point, Equation 2.7 and Equation 2.8 can be recursively used from the top plate 1 to the reboiler (the reboiler can be considered as the (N+1)-th plate), as follows.

- Draw the operating line (Equation 2.7) on the equilibrium curve (Equation 2.8) obtained from the thermodynamic correlations and shown in Figure 2.2.

- Starting at point A, draw a horizontal line to cut the operating curve at point 1. Drop a vertical line through 1 to the operating line at point 2. Proceed in this way until you hit the bottom composition x_B.

- Count the number of stages, i.e., points 2, 4, 6, 8, 10, etc.

This procedure relates the distillate composition x_D to the still composition x_B through the number of stages. In the case of batch distillation, however, the still composition x_B does not remain constant as observed in continuous distillation. This necessitates the use of the recursive scheme several times. If the scheme is used, while keeping the reflux ratio constant throughout the operation just like a normal continuous distillation column, the composition of the distillate keeps changing. On the other hand, the composition of an important key component distillate composition can be maintained as constant by changing the reflux ratio. Therefore, the two basic methods of operating batch distillation columns are:

- Constant reflux and variable product composition

- Variable reflux and constant product composition of the key component

These are described in the following sections.

2.2 Constant Reflux

Smoker and Rose (1940) presented an analysis of the constant reflux operation of batch distillation for the first time in the context of a binary system. They used the Rayleigh equation in conjunction with the McCabe–Thiele method presented above to capture the dynamics of the batch distillation column. In their procedure, the right hand side of the Rayleigh equation (Equation 2.1) is integrated graphically by plotting $\frac{1}{x_D - x_B}$ versus x_B and the area under the curve between the feed composition x_F and the still composition x_B gives the value of the integral.

To establish the relation between the distillate composition x_D and the still composition x_B, the graphical analysis presented in the earlier section is used here. Several values of x_D are selected, and operating lines with the same slope

$\frac{R}{R+1}$ are drawn through the intersection of these x_D values and the diagonal. Once these lines are drawn, steps are drawn between the operating line and the equilibrium curve to get the bottom composition, as shown in Figure 2.3. The operation is stopped once a specific criterion is met, such as the desired average distillate composition in the condenser, or the still composition at the end of the fraction, or the expected total amount of product is collected in the condenser or in the reboiler. Alternatively, one could specify the batch time as a stopping criterion. Although Smoker and Rose presented the calculation method independent of time, time can be introduced through the vapor boilup rate V of the reboiler as follows.

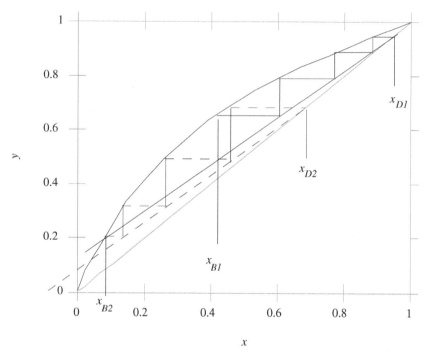

FIGURE 2.3
McCabe–Thiele Method for the Constant Reflux Mode

From the overall material balance it can be seen that the rate of decrease of the still amount, dB/dt, is the same as the distillate rate dD/dt, i.e., $\frac{dB}{dt} = -\frac{dD}{dt}$. Hence, from Equation 2.5 the rate of change of B is given by:

$$\frac{dB}{dt} = -\frac{V}{R+1} \qquad (2.10)$$

By integrating the above equation, we get:

$$\int_0^T dt = \int_B^F \frac{R+1}{V} dB \qquad (2.11)$$

BASIC MODES OF OPERATION

$$T = \frac{R+1}{V}(F - B) = \frac{R+1}{V}D \qquad (2.12)$$

Similarly, one can also obtain estimates of energy requirements from this preliminary analysis. The heat supplied to the reboiler is essentially used for providing the reflux for the column. Therefore, the total heat supplied to the reboiler can be calculated using:

$$Q_R = \int_0^D \lambda R dD = \lambda R D \qquad (2.13)$$

where λ represents the average latent heat of vaporization of the mixture.

Example 2.1: A mixture of components A and B with a 0.40 mole fraction of A is distilled in a batch distillation column containing 5 theoretical stages. The column is operated under the constant reflux condition with a reflux ratio of 5.630 and $\alpha_{a,b} = 2.0$.

a) Find the distillate and the still composition when 40 percent of the mixture is distilled. What is the average distillate composition?

b) If the still contains initially 100 moles of the mixture and the vapor boilup rate is 143.1, what is the total time required to complete the distillation operation?

c) Assume the average latent heat of the mixture to be 40 kJ/g moles and find the energy required to be supplied to the reboiler to complete the above operation.

Solution: The distillation operation requires 40 percent of the mixture to be distilled, which corresponds to 60 percent of the original feed F remaining in the reboiler, i.e., $B = 0.6F$. The problem is that of determining the distillate and still composition, for which the Rayleigh equation (Equation 2.1) applies. To attain the specified requirements, the left hand side of the Rayleigh equation can be calculated as:

$$\ln\left(\frac{F}{B}\right) = \ln\left(\frac{1}{0.6}\right) = 0.5108$$

For various values of x_D the operating lines are drawn and the bottom composition is calculated using the procedure shown in Figure 2.2. By trial and error, we discover that the initial value of x_D corresponding to the still composition of $x_F = 0.4$ is 0.9471. The rest of the x_D are chosen below this composition, and various values of x_B, $x_D - x_B$, $\frac{1}{x_D - x_B}$ are obtained.

Values of x_B versus $\frac{1}{x_D - x_B}$ are plotted in Figure 2.4, from which $\int_{x_F}^{x_B} \frac{dx_B}{x_D - x_B}$ is obtained for each value of x_B. The operation is stopped when the integral is equal to 0.5108.

a) The distillate composition x_D at the end of the operation is 0.5681, and the still composition x_B is 0.1000. The average distillate composition is obtained using the following material balance equations.

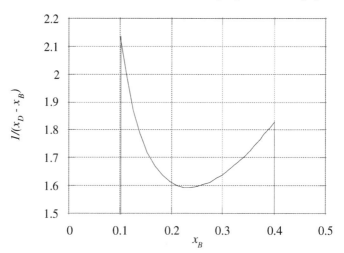

FIGURE 2.4
Graphical Integration for Example 2.1

$$\int_0^D x_D dD = \int_F^B x_B dB$$

$$x_{Dav} \int_0^D dD = Fx_F - Bx_B$$

$$x_{Dav} = \frac{Fx_F - Bx_B}{D}$$

$$x_{Dav} = \frac{100 \times 0.4 - 60 \times 0.1}{40} = 0.85$$

b) The time required for the distillation is given by Equation 2.12:

$$T = \frac{R+1}{V} \times D = 1.8532 \text{ hr}$$

c) To obtain the heat duty of the reboiler, using Equation 2.13:

$$Q_R = 40,000 \times 5.63 \times 40 = 9008 \text{ kJ}$$

2.3   Variable Reflux

In the variable reflux mode of operation, in order to maintain the composition of a key component at a constant[1], the reflux ratio is changed continuously. In 1937, Bogart presented an analysis of variable reflux conditions for a binary system. The steps involved in the calculation procedure are similar to those in the case of the constant reflux mode; however, in the variable reflux case the reflux ratio is varied instead of the distillate composition at each step (see the previous section for details). Moreover, the Rayleigh equation (Equation 2.1), though valid for the variable reflux condition, takes a simplified form, as shown below.

Rayleigh equation (Equation 2.1):

$$\ln\left(\frac{B}{F}\right) = \int_{x_F}^{x_B} \frac{dx_B}{x_D - x_B}$$

Since the distillate composition remains constant (remember that we are considering binary systems here) throughout the operation, the Rayleigh equation reduces to:

$$\frac{B}{F} = \frac{x_D - x_F}{x_D - x_B} \tag{2.14}$$

The second step is to establish the relation between R and x_B. Several values of R are selected, operating lines are drawn through the fixed point (x_D, x_D) with the slope equal to $\frac{R}{R+1}$, and steps are drawn between the operating line and the equilibrium curve (see Figure 2.2 and Figure 2.5) to get the bottom composition. The recursive scheme presented in section 2.1 is repeated until the desired stopping criterion is met. The stopping criterion for the variable reflux case can be still composition at the end of the fraction, or the total amount of product collected in the condenser or reboiler. Similar to the case with constant reflux, the batch time could be a stopping criterion and can be introduced in the calculation procedure as follows.

Integrating the overall material balance equation given by Equation 2.10 results in

$$\int_0^T dt = \int_B^F \frac{R+1}{V} dB \tag{2.15}$$

Obtaining the differential of B with respect to x_B from Equation 2.14 and substituting the results into the above equation, we get

$$T = \int_{x_B}^{x_F} \frac{R+1}{V} \frac{F(x_D - x_F)}{(x_D - x_B)^2} dx_B. \tag{2.16}$$

---

[1] Note that for binary systems this defines the complete composition of the product.

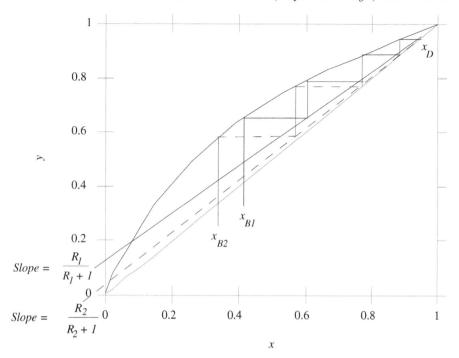

FIGURE 2.5
McCabe–Thiele Method for the Variable Reflux Mode

Similarly, one can also obtain estimates of the heat supplied to the reboiler Q_R from the amount of material refluxed for the given batch, as shown below.

$$Q_R = \int_0^D \lambda R \, dD \qquad (2.17)$$

The graphical integration technique was used earlier in the case of the constant reflux condition to obtain the left hand side of the Rayleigh equation and not to calculate the batch time or the heat requirement. However, since the reflux is variable (i.e., not constant), one has to resort to graphical integration to calculate the batch time and the heat requirements. The Rayleigh equation is straightforward in the case of variable reflux and, therefore, can be integrated without resorting to the graphical procedure. The following example demonstrates the design procedure for the variable reflux condition.

Example 2.2: Rework the problems in Example 2.1 for the variable reflux mode.
Solution: Since the distillate composition is held constant throughout the variable reflux mode of operation, the distillate composition $x_D = x_{Dav} = 0.85$. For the various iterates of R, we obtain the corresponding values of x_B. The value of the amount of product distilled at each x_B is also calculated using the Rayleigh equation for the variable reflux condition (Equation 2.14).

$$D = F\left(1 - \frac{x_D - x_F}{x_D - x_B}\right)$$

a) The operation is stopped when the amount of product collected is greater than or equal to 40. The still composition at $D = 40$ is found to be equal to 0.10.

b) The time required for this operation is calculated by plotting the quantity $\frac{R+1}{V}\frac{F(x_D - x_F)}{(x_D - x_B)^2}$ versus x_B and then finding the area under the curve between x_B equal to 0.4 (Figure 2.6) and 0.1. The time required is found to be 1.995 hrs.

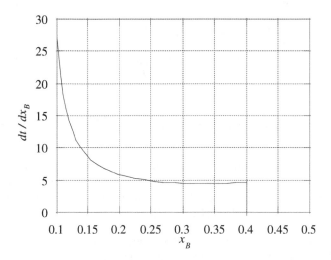

FIGURE 2.6
Graphical Integration for Calculation of Batch Time for Example 2.2

c) Figure 2.7 shows the plot of R versus D, and the area under the curve between D equal to 0 and 40 gives the reboiler heat requirements and is equal to $\lambda \times$ area $= 9800$ kJ.

2.4   Optimal Reflux

From the two examples (Examples 2.1 and 2.2) given above, it is obvious that although both the operations give the same amount of product with the same average product purity, the total time required differs for the two operations. In the case of the constant reflux operation, the product composition is allowed to vary while the reflux is kept constant, whereas in the variable reflux case

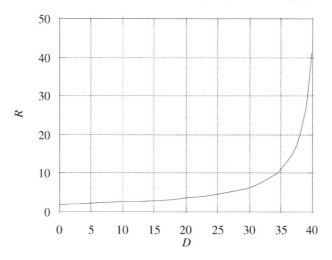

FIGURE 2.7
Graphical Integration for Calculation of Reboiler Heat Duty for Example 2.2

the composition is kept constant by varying the reflux. Let us consider a third mode of operation where neither the reflux nor the product composition is kept constant. For example, consider the following reflux profile for the same separation given in Examples 2.1 and 2.2.

x_D	.9019	.8980	.8899	.8749	.8455	.8265	.8010	.7635	.7018
R	5.472	5.349	5.237	5.138	4.980	5.348	5.966	6.839	7.623

Remember we are using the same batch and the same column for this operation. Since in this case neither distillate composition nor reflux ratio is constant, the following procedure is used for integration of the Rayleigh Equation (Equation 2.1).

For each pair of values of x_D and R, operating lines are drawn with a slope equal to $\frac{R}{R+1}$ and passing through the point (x_D, x_D). The graphical procedure described in Figure 2.2 is used to obtain the corresponding values of x_B. Figure 2.8 outlines this procedure in general.

As stated earlier, the basic batch distillation column satisfies the Rayleigh equation (Equation 2.1). Therefore, we can use the same equation to calculate the total amount of distillate, as was done for the constant reflux condition. Values of x_B, $x_D - x_B$, $1/x_D - x_B$ are obtained for each operating line. Values of x_B versus $1/x_D - x_B$ are plotted in Figure 2.9 and from which the right hand side of the Rayleigh equation is obtained as the area under the curve and is found to be 0.5108, which corresponds to $\ln[\frac{B}{F}]$. Hence the total amount of distillate given by $F(1 - \frac{B}{F})$ is equal to 40 moles, which is what we obtained in the variable reflux and the constant reflux operations in Examples 2.1 and 2.2.

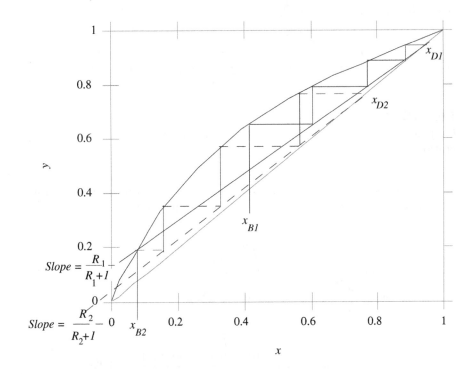

FIGURE 2.8
McCabe–Thiele Procedure for the Third Mode of Operation

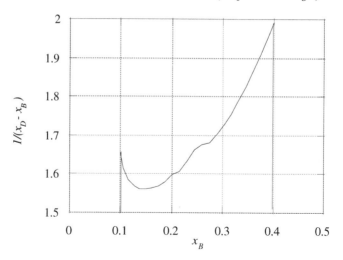

FIGURE 2.9
Graphical Integration for the Rayleigh Equation for the Third Mode of Operation

We resort to the basic mass balance equation to obtain the average distillate composition used in the case of constant reflux mode.

$$\int_0^D x_D dD \; = \; \int_F^B x_B dB$$

$$x_{Dav} \int_0^D dD \; = \; F x_F \; - \; B x_B$$

$$x_{Dav} \; = \; \frac{F x_F \; - \; B x_B}{D}$$

$$x_{Dav} \; = \; \frac{100 \times 0.4 \; - \; 60 \times 0.1}{40} \; = \; 0.85$$

The average distillate composition in this case is found to be 0.85, which is again the same as that of the other two operating modes.

Similarly, the time requirement for this operation is derived from the basic material balance equations as was done in the case with the variable reflux condition. Integrate the overall material balance equation given by Equation 2.10 as follows:

$$\int_0^T dt \; = \; \int_B^F \frac{R+1}{V} dB \qquad (2.18)$$

Use the Rayleigh equation in the following form to obtain the value of dB to be substituted in the above equation.

$$dB = B\frac{dx_B}{x_D - x_B} \tag{2.19}$$

Using the above two equations, the time T required when neither distillate composition nor reflux is constant is found to be

$$T = \int_{x_B}^{x_F} \frac{B}{V}\frac{R+1}{x_D - x_B}dx_B \tag{2.20}$$

So the time required for the above operation is obtained by the graphical integration shown in Figure 2.10, where $\frac{B}{V}\frac{R+1}{x_D - x_B}$ versus x_B is plotted. The value of T is found to be 1.686 hr, smaller than for the constant and variable reflux modes of operation.

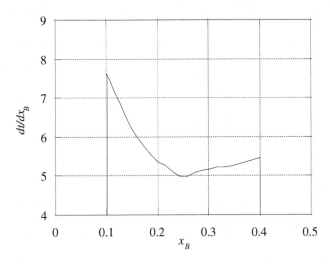

FIGURE 2.10
Graphical Integration for Calculation of Batch Time for the Third Mode of Operation

It can be seen that this reflux profile requires the least amount of time to obtain the same amount of product with the same purity. This policy is neither variable reflux nor constant reflux but is in between the two. The policy which was specified in the above example is the optimal policy obtained by Coward in 1967. This type of operation is possible in batch distillation and is known as the optimal reflux or the optimal control policy.

The *optimal reflux policy* is essentially a *trade-off* between the two operating modes of constant reflux and variable reflux, and is based on the ability to yield the most profitable operation. The calculation of this policy is a difficult problem and relies on optimal control theory. A separate chapter is dedicated to discussing the solution procedures for this operating mode.

The problem we have seen in this chapter is called the minimum time problem. There are different kinds of optimal reflux policies depending on the indices of performance chosen as the objective. The indices generally used in practice include minimum time, maximum distillate, and maximum profit function.

References

Bogart M. J. P. (1937), The design of equipment for fractional batch distillation, *Transactions of American Institute of Chemical Engineers*, **33**, 139.

Bowman J. R. and M. T. Cichelli (1949), Batch distillation: Minimum number of plates and minimum reflux, *Industrial and Engineering Chemistry*, **41**, 1985.

Converse A. O. and G. D. Gross (1963), Optimal distillate-rate policy in batch distillation, *Industrial and Engineering Chemistry Fundamentals*, **2**, 217.

Coward I. (1967), The time optimal problem in binary batch distillation, *Chemical Engineering Science*, **22**, 503.

McCabe W. L. and E. W. Thiele (1925), Graphical design of fractionating columns, *Industrial and Engineering Chemistry*, **17**, 605.

Smoker E. H. and A. Rose (1940), Graphical determination of batch distillation curves for binary mixtures, *Transactions of American Institute of Chemical Engineers*, **36**, 285.

Exercises

2.1 A mixture containing 48.6 percent benzene and 51.4 percent ethylene chloride is to be distilled in a batch distillation column with 20 theoretical plates. The column is operating under the constant reflux mode at a reflux ratio equal to 9.8. Assume the relative volatility of benzene with respect to ethylene chloride is 1.109.

a) Find the concentration of the product when 95.4 percent of the total mixture is distilled. Assume that the initial distillate composition is approximately 0.722.

b) If the original feed is 100 moles and the vapor boil-up rate is 100 moles/hr, what is the time required to complete the operation?

c) Simulate the variable reflux operation assuming the reflux ratio varies from 5.0 to 20.0. Find the end compositions and the total time.

2.2 A batch distillation column operating at atmospheric pressure is to be designed to separate a mixture containing 15.67 mole percent CS_2 and 84.33 mole percent CCl_4 into an overhead product containing 91 percent CS_2. Assume the column to be operating in the variable reflux mode with an initial reflux ratio of 3.0.

a) How many theoretical plates are required for the process?

b) If the distillate is stopped when the reflux ratio is equal to 15.0, what is the amount of distillate obtained?

c) What is the heat required per kmole of product?

Latent heat for the CS_2 and CCl_4 mixture is 25,900 kJ/kmol and the data for the equilibrium curve is given below.

x_{CS2} .0296 .0615 .1106 .1435 .2585 .3900 .5318 .6630 .7575 .8604
y_{CS2} .0823 .1555 .2660 .3320 .4950 .6340 .7470 .8290 .8780 .9320

2.3 If the same system is operating with the constant reflux mode of operation, with the initial product composition of 95 percent and with a reflux ratio of 5.0,

a) How many theoretical plates are required?

b) Simulate the condition and stop the operation when the average distillate composition is 0.90. What is the amount of distillate collected at this stage? What is the batch time if $V = 100$ moles?

c) Find the heat required per kmole of product.

2.4 A multicomponent mixture containing meta-, ortho-, and para-mono-nitro-toluene is to be distilled using a distillation column containing eight theoretical plates. The feed composition at the start of operation is 0.6, 0.36, 0.04 of meta-, ortho-, and para-mono-nitro-toluene respectively. The column is operating at a reflux ratio equal to 3.0 and with the constant reflux mode of operation. Assume that the product composition varies from 0.94 to 0.85. The relative volatilities of meta-, ortho-, and para-mono-nitro-toluene can be assumed as 1.7, 1.16, and 1.0, respectively. Use the plate-to-plate calculation method to calculate the relation between the distillate composition and the still composition at any instant.

a) Plot the distillate compositions of meta-, ortho-, and para-mono-nitro-toluene versus the still composition of meta-mono-nitro-toluene.

b) Plot the still compositions of ortho- and para-moono-nitro-toluene versus the still composition of meta-mono-nitro-toluene.

c) Find the fraction distilled at the end of the operation.

2.5 The same column in Example 2.4 is operated with the variable reflux mode of operation. The distillate purity of the meta-mono-nitro-toluene is to be maintained at 0.98. Assume that the reflux ratio varies from 10 to 80.

a) Plot the distillate compositions of ortho- and para-mono-nitro-toluene versus the still composition of meta-mono-nitro-toluene.

b) Plot the still compositions of ortho- and para-mono-nitro-toluene versus the still composition of meta-mono-nitro-toluene.

c) Plot reflux ratio versus the still composition of meta-mono-nitro-toluene.

d) Find the fraction distilled at the end of the operation.

2.6 Converse and Gross, in 1963, solved the maximum distillate optimal reflux problem given below (Converse and Gross, 1963).

> Maximum Distillate Problem – Maximize the amount of distillate of a specified concentration for a specified time.

Use their system to compare the three modes of operation.

2.7 Find the heat duty of the reboiler for the minimum time problem solved in the section on optimal reflux (Section 2.4).

2.8 Bowman and Cichelli, in 1949, presented a very interesting concept of pole height for a binary batch distillation column. A pole height is defined as the product of the mid-point of the slope of the distillate composition versus material remaining in the still curve and the amount of the material remaining in the still at that time. Figure 2.11 illustrates the concept of pole height. They stated that the

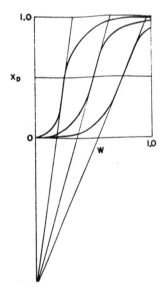

FIGURE 2.11
The Pole Height Concept (Reproduced from Bowman and Cichelli, 1948)

pole height is invariant to the initial concentration and provides a good measure for defining sharpness of separation.

Take 100 moles of a binary mixture containing component A & B with a relative volatility of 1.5. Use a five theoretical stage batch distillation column and a constant reflux operation with a reflux equal to 5.0. Vary the initial composition from x_A equal to 0.5 to 0.4 and plot the distillate composition versus the amount remaining in the still. Calculate the pole height for each case. Verify the concept.

2.9 Problem 2.8 described the pole height concept proposed by Bowman and Cichelli. Use the above definition to prove that:

a) At a total reflux condition (minimum number of plates), the pole height S is related to the number of plates N by the following relation.

$$S = \frac{\alpha^N}{8F(1 - x_F)}$$

b) At an infinite number of plates (minimum reflux condition) and moderately good separation, the reflux can be expressed as:

$$R = \frac{2S - \alpha}{\alpha - 1}$$

Assume that at moderate separation the quantity of x_B is extremely small.

3

COLUMN DYNAMICS

CONTENTS

In Chapter 2, we examined the different modes of batch distillation operation. The theoretical analysis and a graphical approach based on simplified assumptions were developed for the different operating modes, and the approach was illustrated using binary systems. This chapter presents a rigorous analysis of the batch distillation column. It begins with a description of the dynamics of the column in terms of differential mass and energy balances. This is followed by a general discussion on numerical integration techniques. Finally, the errors and stability issues of the governing equations are analyzed, solution techniques in the context of batch column dynamics are outlined, and a few simulation examples are given.

3.1 Governing Equations

The early models (Huckaba and Danly, 1960) that considered the complete dynamics of batch distillation were restricted to ideal binary systems. In 1963, Meadows presented the first comprehensive model of batch distillation for the constant reflux mode of operation. However, detailed analysis of the characteristics of differential mass and energy balances associated with the complete dynamics of a multicomponent batch distillation column was presented for the first time by Distefano (1968a, 1968b). Distefano pointed out that the system of equations presented for batch distillation is much more difficult to solve than that for the dynamics of a continuous column because of several factors.

For example, in the case of batch distillation plate holdups are generally much smaller than the reboiler holdup, while in continuous distillation the ratio of reboiler to plate holdup is not nearly as great. Also, in batch distillation severe transients can occur, unlike the continuous distillation where the variations are relatively small. Distefano's work forms the basis for almost all of the later work on rigorous modeling of batch distillation columns (Boston et al., 1983; Diwekar, 1988; Diwekar and Madhavan, 1991). Therefore, the governing equations and part of the analysis presented in this chapter follow the original developments put forth by Distefano.

The model presented in this section is based on the assumptions of negligible vapor holdup, adiabatic operation, and theoretical plates.

Figure 3.1 represents a schematic of a batch distillation column, where the holdup on each plate is responsible for the dynamics of each plate. For an arbitrary plate j, the total mass, component, and energy balances yield the following equations.

Overall material balance around plate j:

$$\frac{dH_j}{dt} = V_{j+1} + L_{j-1} - V_j - L_j \tag{3.1}$$

Component balance for the component i around plate j:

$$\frac{dH_j x_j^{(i)}}{dt} = V_{j+1} y_{j+1}^{(i)} + L_{j-1} x_{j-1}^{(i)} - V_j y_j^{(i)} - L_j x_j^{(i)} \tag{3.2}$$

Energy balance around plate j:

$$\frac{dH_j I_j}{dt} = V_{j+1} J_{j+1} + L_{j-1} I_{j-1} - V_j J_j - L_j I_j \tag{3.3}$$

where $j = 0, 1, \ldots, N$ and $i = 1, 2, \ldots, n$, and

- H_j is the molar holdup on plate j [mole]

- L_j is the liquid stream leaving plate j [mole/hr]

- n is the number of components

- N is the number of plates

- V_j is the vapor stream leaving plate j [mole/hr]

- $x_j^{(i)}$ is the liquid composition of the component i on plate j [mole fraction]

- $y_j^{(i)}$ is the vapor composition of the component i which is in equilibrium with the liquid on plate j [mole fraction]

- I_j is the enthalpy of the liquid stream leaving plate j [J/mole]

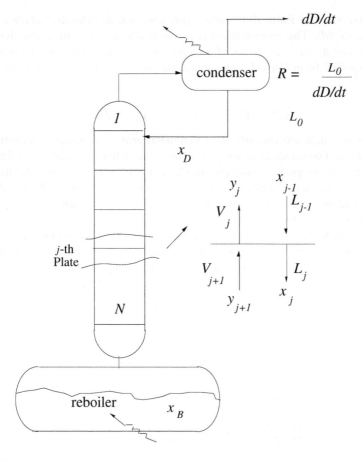

FIGURE 3.1
Schematic of a Batch Distillation Column

- J_j is the enthalpy of the vapor stream leaving plate j [J/mole]

Substitution of Equation 3.1 in Equations 3.2 and 3.3 results in the following equations:

$$\frac{dx_j^{(i)}}{dt} = \frac{1}{H_j}[V_{j+1}(y_{j+1}^{(i)} - x_j^{(i)}) + L_{j-1}(x_{j-1}^{(i)} - x_j^{(i)}) - V_j(y_j^{(i)} - x_j^{(i)})] \quad (3.4)$$

and

$$V_{j+1} = \frac{1}{J_{j+1} - I_j}[V_j(J_j - I_j) + L_{j-1}(I_j - I_{j-1}) + H_j\delta I_j] \quad (3.5)$$

Equation 3.5 has been rearranged as shown above so that the last term,

δI_j, represents the finite difference approximation for the derivative of the enthalpy, dI_j/dt. This approximation is the result of the assumption that the flowrates are much larger than the change in enthalpy. Similar assumptions can be applied to holdup changes in Equation 3.1, resulting in the following equation[1]:

$$L_j = V_{j+1} + L_{j-1} - V_j - \delta H_j \qquad (3.6)$$

These assumptions simplify the calculation procedure without compromising accuracy. The approximations could be relaxed in cases where the changes in enthalpy are large. For example, in the study of reactive batch distillation (Cullie and Reklaitis, 1986) the enthalpy changes were found to be dominant and the difference equations for the enthalpy balance (Equation 3.5) were replaced by differential equations.

Similarly, for the top of the column the material balance and energy balance for the condenser and the accumulator system are represented by the following equations, given that the reflux ratio R is related to the distillate rate dD/dt by $L_0 = R \, dD/dt$ and the condenser heat duty is Q_D.

$$L_0 = R\frac{dD}{dt} \qquad (3.7)$$

$$\frac{dH_D}{dt} = V_1 - L_0 - \frac{dD}{dt} = V_1 - \frac{dD}{dt}(R+1) \qquad (3.8)$$

Mass balance for component i is given by:

$$\frac{dH_D x_D^{(i)}}{dt} = V_1 y_1^{(i)} - L_0 x_D^{(i)} - x_D^{(i)}\frac{dD}{dt} \qquad (3.9)$$

The energy balance equation is

$$\frac{dH_D I_D}{dt} = V_1 J_1 - L_0 I_D - dD/dt \, I_D - Q_D \qquad (3.10)$$

where

- $\frac{dD}{dt}$ is the distillate rate [mole/hr]

- $H_D = H_0$ is the molar condenser holdup [mole]

- I_D is the enthalpy of the liquid in the condenser [J/mole]

- L_0 is the liquid reflux at the top of the column [mole/hr]

---

[1]In the context of continuous distillation, it is well established that the assumptions about constant molar or constant volume or constant weight holdup yield similar results. We extend these observations to batch distillation, and because the balances are generally expressed in moles rather than volume or weight, the constant molar holdup can be used here for simplicity. The dynamics of batch distillation in Table 3.1 use this assumption.

- R is the reflux ratio

- $x_D^{(i)} = x_0^{(i)}$ is the composition of component i in the distillate [mole fraction]

Applying the same finite difference assumptions for the change in holdup and change in enthalpy, Equations 3.8 to 3.10 transform into the following equations.

The vapor flowrate at the top of the column:

$$V_1 = (R+1)\frac{dD}{dt} + \delta_t H_D \qquad (3.11)$$

The differential balance for the distillate composition of component i is

$$\frac{dx_D^{(i)}}{dt} = \frac{V_1}{H_D}(y_1^{(i)} - x_D^{(i)}) \qquad (3.12)$$

Substituting Equation 3.11 in Equation 3.10, the condenser heat duty Q_D is

$$Q_D = V_1(J_1 - I_D) - H_D\delta_t I_D \qquad (3.13)$$

An analogous procedure for the reboiler results in the following reboiler equations. Here the holdup in the reboiler is equivalent to the amount remaining in still B.

The overall material balance around the reboiler is

$$\frac{dB}{dt} = L_N - V_B \qquad (3.14)$$

The differential balance for the still composition of component i is

$$\frac{dx_B^{(i)}}{dt} = \frac{1}{B}[L_N(x_N^{(i)} - x_B^{(i)}) - V_B(y_B^{(i)} - x_B^{(i)})] \qquad (3.15)$$

The reboiler heat duty Q_B is given by:

$$Q_B = V_B(J_B - I_B) - L_N(I_N - I_B) + B\delta_t I_B \qquad (3.16)$$

where

- B is the amount of material remaining in the reboiler [mole]

- I_B is the enthalpy of the liquid in the reboiler [J/mole]

- V_B is the vapor stream leaving the reboiler [mole/hr]

- $x_B^{(i)}$ is the liquid composition of component i in the reboiler [mole fraction]

- $y_B^{(i)}$ is the vapor composition of component i which is in equilibrium in the reboiler [mole fraction]

In short, differential Equations 3.4, 3.12, 3.14, and 3.15 describe the component composition changes and the change in still amount. Algebraic Equations 3.5, 3.6, and 3.11 are used to calculate the vapor and liquid flowrates at each plate. The heat duties are computed using Equations 3.13 and 3.16. The thermodynamics provide the equilibrium and the enthalpy relations for each plate and for the condenser and reboiler [2].

Thermodynamic correlations:

$$y_j^{(i)} = f((x_j^{(k)}, k = 1, \ldots, n), TE_j, P_j) \qquad (3.17)$$

Enthalpy relations:

$$I_j = f((x_j^{(k)}, j = 1, \ldots, n), TE_j, P_j)$$
$$J_j = f((y_j^{(k)}, j = 1, \ldots, n), TE_j, P_j) \qquad (3.18)$$

Table 3.1 lists all these equations involved in the dynamic analysis of the batch column and the assumptions behind these equations.

The system shown in Table 3.1 represents a generalized form of the batch distillation column. The treatment of individual modes of operation, such as constant reflux and variable reflux, is presented in the section 3.4 on solution strategies. Furthermore, the initial conditions for these equations are obtained from the initial total reflux operation. This initial total reflux startup operation for batch distillation attains a steady state condition in an equilibration time of T_{eq}. The transient analysis of the total reflux condition is also described later in this chapter, after the discussion on stability analysis and numerical integration techniques. Alternatively, one can use direct steady state analysis at equilibration time to generate the initial conditions for the complete column dynamics shown in Table 3.1. Table 3.2 shows the algebraic equations at the end of the total reflux operation which need to be solved iteratively to generate the initial conditions.

One can easily see from the system of differential equations shown in Table 3.1 that there is no analytical solution to the problem and one must resort to numerical solution techniques. Two main problems arise in the numerical solution of differential equations: truncation error and numerical instability. The following section describes the different integration techniques and their error and stability characteristics. Most of the discussion in Section 3.2 below is derived from Distefano (1968b), Fatunla (1988), and Golub and Ortega (1992).

---

[2]The hydraulic models for calculation of holdups and pressure drop across plates are not presented in the context of batch distillation so far. A preliminary analysis of hydraulic models was presented in the context of variable reflux operation but this area needs more investigation. Hence these models are not included in the rigorous model presented here.

TABLE 3.1
Complete Column Dynamics

Assumptions
• Negligible vapor holdup,
• Adiabatic operation,
• Theoretical plates,
• Constant molar holdup,
• Finite difference approximations for the enthalpy changes.

Composition Calculations
Condenser and Accumulator Dynamics
$$\frac{dx_D^{(i)}}{dt} = \frac{V_1}{H_D}(y_1^{(i)} - x_D^{(i)}), i = 1,2,\ldots,n$$
Plate Dynamics
$$\frac{dx_j^{(i)}}{dt} = \frac{1}{H_j}[V_{j+1}y_{j+1}^{(i)} + L_{j-1}x_{j-1}^{(i)} - V_j y_j^{(i)} - L_j x_j^{(i)}],$$
$$i = 1,2,\ldots,n; j = 1,2,\ldots,N$$
Reboiler Dynamics
$$\frac{dx_B^{(i)}}{dt} = \frac{1}{B}[L_N(x_N^{(i)} - x_B^{(i)}) - V_B(y_B^{(i)} - x_B^{(i)})], i = 1,2,\ldots,n$$

Flowrate Calculations
At the Top of the Column
$$L_0 = R\frac{dD}{dt}; V_1 = (R+1)\frac{dD}{dt}$$
On the Plates
$$L_j = V_{j+1} + L_{j-1} - V_j; j = 1,2,\ldots,N$$
$$V_{j+1} = \frac{1}{J_{j+1} - I_j}[V_j(J_j - I_j) + L_{j-1}(I_j - I_{j-1}) + H_j\delta I_j]$$
$$j = 1,2,\ldots,N$$
At the Bottom of the Column
$$\frac{dB}{dt} = L_N - V_B$$

Heat Duty Calculations
Condenser Duty
$$Q_D = V_1(J_1 - I_D) - H_D\delta_t I_D$$
Reboiler Duty
$$Q_B = V_B(J_B - I_B) - L_N(I_N - I_B) + B\delta_t I_B$$

Thermodynamics Models
Equilibrium Relations
$$y_j^{(i)} = f((x_j^{(k)}, k=1,\ldots,n), TE_j, P_j)$$
Enthalpy Calculations
$$I_j = f((x_j^{(k)}, j=1,\ldots,n), TE_j, P_j)$$
$$J_j = f((y_j^{(k)}, j=1,\ldots,n), TE_j, P_j)$$

TABLE 3.2

Steady State at the End of the Total Reflux Operation

Composition Calculations
Condenser Balance
$0 = (y_1^{(i)} - x_D^{(i)}), i = 1, 2, \ldots, n$
Plate-to-Plate Calculations
$0 = [V_{j+1} y_{j+1}^{(i)} + L_{j-1} x_{j-1}^{(i)} - V_j y_j^{(i)} - L_j x_j^{(i)}],$ $i = 1, 2, \ldots, n; j = 1, 2, \ldots, N$
Reboiler Balance
$0 = [L_N(x_N^{(i)} - x_B^{(i)}) - V_B(y_B^{(i)} - x_B^{(i)})]$
Overall Component Balance
$Fx_F^{(i)} = [H_D x_D^{(i)} + \sum_{j=1}^{N} H_j x_j^{(i)} + Bx_B^{(i)}]$ $i = 1, 2, \ldots, n$
Flowrate Calculations
At the Top of the Column
$V_1 = L_0$
On the Plates
$L_j = V_{j+1} + L_{j-1} - V_j; j = 1, 2, \ldots, N$
$V_{j+1} = \frac{1}{J_{j+1} - I_j}[V_j(J_j - I_j) + L_{j-1}(I_j - I_{j-1})$
Thermodynamics Correlations
Equilibrium Relations
$y_j^{(i)} = f(x_j^{(k)}, k = 1, \ldots, n, TE_j, P_J)$
Enthalpy Calculations
$I_j = f((x_j^{(k)}, j = 1, \ldots, n), TE_j, P_j)$
$J_j = f((y_j^{(k)}, j = 1, \ldots, n), TE_j, P_j)$

3.2 Error and Stability

Numerical solution techniques for differential equations normally involve approximating the differential equations by difference equations that are solved in a step-by-step marching fashion. A major question regarding any numerical integration technique is the accuracy of the approximation, which depends upon the errors, convergence, and stability of the integration techniques.

For example, consider the following system of equations:

$$\frac{d\bar{y}(\bar{x})}{d\bar{x}} = f(\bar{x}, \bar{y}(\bar{x})) \qquad (3.19)$$

Here $\bar{y}$ and $\bar{x}$ denote the vector of variables. Any numerical integration scheme for the equation shown above can be represented in the following general form where h denotes the step size and $k+1$ represents the next step:

$$y_{k+1} = \sum_{i=1}^{m} \beta_i y_{k+1-i} + h\phi(x_{k+1}, \ldots, x_{k+1-m}, x_{k+1}, \ldots, x_{k+1-m}) \qquad (3.20)$$

where m is the order of the polynomial and ϕ is some functional.

In general, the error in applying these techniques comes from two sources: the truncation error that results from the replacement of the differential equation by the difference approximation (η) and the rounding error (ϵ) made in carrying out the arithmetic operations of the method. The above approximation can be rewritten using the error terms as follows.

Local truncation error:

$$y(x_{k+1}) - y_{k+1} = h\eta_k \qquad (3.21)$$

where $y(x_{k+1})$ is the true solution and y_{k+1} is the finite-difference approximation.

Rounding errors:

$$y_{k+1} = \sum_{i=1}^{m} \beta_i y_{k+1-i} + h\phi(x_{k+1}, \ldots, x_{k+1-m}, x_{k+1}, \ldots, x_{k+1-m}, \eta_k^1) + \epsilon_k^2 \qquad (3.22)$$

Normally, the truncation error goes to zero as h approaches zero. Hence, we can make the error as small as we want by choosing an appropriate h. However, the smaller the h, the larger the rounding error on the computed solution. In practice, for a fixed word length in computer arithmetic, there will be a size h below which the rounding error will become the dominant contribution to the overall error. Because a faster solution is always preferred, it is desirable to choose as big a step size as possible where the truncation error is predominant.

For a method to be convergent, the finite difference solution must approach the true solution as the interval or the step size approaches zero.

The concept of stability is associated with the propagation of errors of the numerical integration technique as the calculation progresses with a finite interval size and is related to the effects of errors made in a single step on succeeding steps. The problem of stability arises because in most instances the order m of the approximate difference equation (Equation 3.20) is higher than the order q of the original differential equation (Equation 3.19), whose actual solution can be written as $(m > q)$:

$$y(x) \;=\; \sum_{i=1}^{q} \gamma_i g_i(\bar{x}) \tag{3.23}$$

Hence, the difference equation may contain extraneous terms that may dominate the equation and bear little or no resemblance to the true solution. It happens frequently that the spurious solutions do not vanish even in the limit as the increment size approaches zero. This phenomenon is called *strong instability* and implies lack of convergence as well as lack of stability. When a method possesses convergence but has unstable asymptotic behavior, the phenomenon is called *weak instability*. For example, a numerical routine that is stable for some finite increment size, such as h_1, but is unstable for some larger increment size, e.g., h_2 $(h_2 > h_1)$, is said to possess a weak instability (Distefano, 1968b). Even if the order of the approximate difference equation is less than or equal to the order of the original differential equation (e.g., lower order methods, such as Euler's integration, which is discussed in the next subsection), the method can be stable over a very small interval. On a finite interval, a stable method gives accurate results for a sufficiently small h. On the other hand, even strongly stable methods can exhibit unstable behavior if h is too large. Although in principle h can be taken to a sufficiently small value to overcome this difficulty, it may be that the computing time then becomes prohibitive. Furthermore, the rounding error can become dominant, resulting in an erroneous solution. This is the situation with differential equations that are known as 'stiff' and is often a feature of batch distillation dynamics. In the next section, we will discuss different numerical integration techniques, their error and stability characteristics, and their applicability to the solution of batch distillation dynamics.

3.3   Numerical Integration Techniques

Numerical integration techniques can be categorized into two types:

- One-Step Methods
- Multistep Methods

Another way of classifying the integration techniques depends on whether or not the method is explicit, semi-implicit, or implicit. The implicit and semi-implicit methods play an important role in the numerical solution of stiff differential equations. To maintain the continuity of the section, we will first describe the explicit integration techniques in the context of one-step and multistep methods. The concept of stiffness and implicit methods are considered in a separate subsection, which also marks the end of this section.

3.3.1   One-Step Methods

In single-step methods, at each step of integration the solution depends on the prior step only. The generalized form of a single-step method, as derived from Equation 3.20, can be written as

$$y_{k+1} \; = \; y_k \; + \; h\phi(x_k, y_k) \tag{3.24}$$

The functional form of ϕ changes according to different methods. The most commonly used methods in this category are Euler's method and Runge–Kutta methods. These methods are described below.

Euler's Method

Euler's method perhaps represents the simplest numerical integration technique, which can be defined as

$$y_{k+1} \; = \; y_k \; + \; hf(x_k, y_k) \tag{3.25}$$

Euler's method is derived from a first order Taylor series approximation, where the higher derivative terms starting from the second derivative are ignored. It can be easily proved that the local truncation error $L(h)$ for Euler's method decreases proportionally to the step size, i.e., $L(h) \; = \; O(h)$.

Runge–Kutta Methods

In Runge–Kutta methods, Euler's simple formula for calculating the functional derivative at one point and then using it to calculate the Taylor series expansion shown in Equation 3.25 is replaced by derivative information at a number of points. As the number of points increases, the order of the method increases. The order of the method is defined to be the integer p for which $L(h) \; = \; O(h^p)$. In other words, for the p-th order method, we expect the local truncation error to decrease by a factor of about 2^p when we halve h, assuming that h is sufficiently small. So, for the fourth-order Runge–Kutta method in the following equation, the local truncation error has the form given by $L(h) \; = \; O(h^4)$.

Fourth-Order Runge–Kutta Method:

$$
\begin{aligned}
k_0 &= hf\left(x_k, y_k\right) \\
k_1 &= hf\left(x_k, y_k + \frac{k_0}{2}\right) \\
k_2 &= hf\left(x_k, y_k + \frac{k_1}{2}\right) \\
k_3 &= hf\left(x_k, y_k + k_2\right) \\
y_{k+1} &= y_k + \frac{1}{6}\left(k_0 + 2k_1 + 2k_2 + k_3\right) \quad\quad (3.26)
\end{aligned}
$$

Higher-order Runge–Kutta methods have a better stability range than Euler's integration technique.

The one-step methods described above are the Taylor series approximation where the higher-order derivative terms are ignored. However, many times this approximation may not hold true, and variable step size methods are then preferable. In spaces where higher-order derivatives cannot be ignored for that step, use very small h, otherwise, use a larger step size. Several different variants of the methods presented here are available where higher-order derivative information can be used to obtain automatic step size changes. For more details, refer to Henrici (1962), Butcher (1987), Hairer et al. (1987), and Golub and Ortega (1992).

3.3.2 Multistep Methods

In one-step methods, the integration formula depends on previous steps (i.e., the $k+1$-th point is calculated from k-th point). However, since all these methods are dependent on polynomial approximations, the information about other points such as $k-1$, $k-2$, etc., is also included in the approximation, which makes these multistep methods more realistic. The multistep methods are dependent on more steps than just the previous one. The multistep methods are attractive because they provide better representation of the functional space and, hence, better accuracy. The following equation provides the generalized representation of the multistep methods where the function f is replaced by a polynomial function p.

$$
y_{k+1} = y_k + \int_{x_k}^{x_{k+1}} f(x, y(x))dx = y_k + \int_{x_{k-j}}^{x_{k+1}} p(x)dx \quad\quad (3.27)
$$

However, multistep methods suffer from two problems not encountered in one-step methods. One problem is associated with the starting of these methods. The $k+1$-th step is dependent on the k and $k-1$ steps, etc., and at $k=1$ there is no information about these previous steps. To circumvent this problem, a one-step method is used as a starter for a multistep method, until all the information is gathered. Alternatively, one may use a one-step method at the first step, a two-step method at the second, and so on, until starting values have been built up. However, it is important that the starting method used maintains the same accuracy in the initial stages as the multistep method (which means that, initially, one must use smaller step sizes).

The second problem with multistep methods is the presence of extraneous solutions. This comes under the category of techniques where the order of the solution method is more than the order of differential equations $(m > q)$. Therefore, while these methods provide greater accuracy, they may also possess strong instability characteristics. The following paragraphs describe some of the multistep methods.

Adams–Bashforth Methods

Euler's method can be considered as the simplest form of the Adams–Bashforth method, where the function f is represented by a polynomial p of order 1. If p is assumed to be a linear function that interpolates between (x_{k-1}, f_{k-1}) and (x_k, f_k), then p can be represented by:

$$p(x) = p_1(x) = f_k - \frac{x - x_k}{h} \delta f_k \qquad (3.28)$$

where $\delta f_k = f_{k-1} - f_k$ and $h = |x_{k-1} - x_k|$. Integrating the above equation and substituting it in Equation 3.27 results in the two-step Adams–Bashforth method formula given below.

$$y_{k+1} = y_k + h f_k - \frac{h}{2} \delta f_k = y_k + \frac{h}{2}(3 f_k - f_{k-1}) \qquad (3.29)$$

Similarly, one can obtain higher-order Adams–Bashforth methods by using higher-order polynomials.

The truncation error for a p-th order multistep method is same as that obtained by a p-th order one-step method, i.e., $L(h) = O(h^p)$. However, in multistep methods higher-order methods can be constructed by evaluating f once per step.

The Adams–Bashforth methods use information about prior points. In principle, one can form polynomials using forward points as well. Using the points $x_{k+1}, x_k, \ldots, x_{k_N}$ to form an $N + 1$ polynomial generates a class of methods known as Adams-Moulton methods. However, in these methods also calculation of y_{k+1} requires the solution of f_{k+1} implicitly. Implicit methods are discussed separately in a section which deals with stiff equations. One can also use a combination of an implicit method, such as an Adams–Moulton method, along with an explicit method, like an Adams–Bashforth method, to form an explicit method known as the Predictor-Corrector method, which is discussed below.

Predictor-Corrector Method

The most commonly used Predictor-Corrector method is the combination of the fourth-order Adams–Moulton method and the fourth-order Adams–Bashforth Method.

At first, the Adams–Bashforth formula is used to calculate the predicted value of y_{k+1}.

Predict:

$$y_{k+1}^p = y_k + \frac{h}{24}(55f_k - 59f_{k-1} + 37f_{k-2} - 9f_{k-3}) \qquad (3.30)$$

This predicted value is then used to calculate f_{k+1}.

Modify:

$$f_{k+1}^p = f(x_{k+1}, y_{k+1}^p) \qquad (3.31)$$

The value of f_{k+1}^p is then substituted in the implicit Adams–Moulton formula to correct the value of y_{k+1}.

Correct:

$$y_{k+1} = y_k + \frac{h}{24}(9f_{k+1}^p + 19f_k - 5f_{k-1} + f_{k-2}) \qquad (3.32)$$

3.3.3   Stiff Equations and Implicit Methods

The concept of stiffness is related to stability and can be understood in the context of the following simple problem from Golub and Ortega (1992).

Consider the differential equation

$$\frac{dy}{dx} = -100y + 100, \quad y(0) = 2.0 \qquad (3.33)$$

The exact solution to this problem is given by

$$y(x) = e^{-100x} + 1 \qquad (3.34)$$

If we apply Euler's method to this problem recursively, we can arrive at the following approximation to the exact solution (Equation 3.34).

$$y_n = (1 - 100h)^n + 1 \qquad (3.35)$$

The limiting value of y is 1, which can be obtained as x tends to ∞. The value of y decreases very rapidly until x is equal to 0.1 and then slowly approaches 1. Therefore, we would expect to obtain sufficient accuracy with Euler's method using a relatively large h. However, if one looks at the exact solution and Euler's approximation using h greater than 0.02, the approximation grows rapidly, making the system unstable. This is because the quantity $(1 - 100h)^n$ is an approximation to e^{-100x} and is, indeed, a good approximation for a small h. Even though this exponential term contributes virtually nothing to the solution after $x = 0.1$, Euler's method still requires that we approximate it with sufficient accuracy to maintain stability. This is the typical problem with stiff equations: the solution contains a component that contributes very little to the solution, but the usual methods require it be approximated accurately.

Stiff initial value problems were first encountered in the study of the motion of springs of varying stiffness, from which the problem derives its name. For linear ordinary differential equations, the stiffness of the system can be defined in terms of the stiffness ratio SR (Finlayson, 1980), given by:

$$SR = \frac{\max |\text{Re}\ (\lambda_i)|}{\min |\text{Re}\ (\lambda_i)|} \qquad (3.36)$$

where $\text{Re}\ (\lambda_i)$ is the real part of an eigenvalue λ_i of the Jacobian matrix $[df/dy]$ for the set of ordinary differential equations defined as $y' = f(y)$. Typically, SR of the order of 10 is considered to be not stiff; SR around 10^3 is stiff; and SR around 10^6 is very stiff. For nonlinear ordinary differential equations, the eigenvalues correspond to the eigenvalues of the Jacobian matrix at that particular time step. The stiffness ratio then applies only to that particular time and may change as the integration proceeds.

Cameron et al. (1986) proposed a different measure to quantify stiffness, called the computational stiffness, $S_c(t)$, defined as:

$$S_c(t) = \max_i \ [\text{Re}(-\lambda_i h)] \qquad (3.37)$$

where h is the step size that accuracy requirements would allow if stability is not a limitation for the method applied to the problem. This stiffness quantifying measure is dependent on the type and order of the numerical integration method used, unlike the earlier measure, i.e., the stiffness ratio. Cameron et al. stated that this measure is much more stringent in identifying stiff systems than the stiffness ratio measure. If the value of computational stiffness is less than 1, then the system is computationally nonstiff. This may happen for small h even if the λ_i is very large, resulting in a large stiffness ratio, indicating the system to be stiff. Since normally the computational burden of stiff algorithms is more than nonstiff algorithms, it is better to switch to nonstiff algorithms whenever possible. The quantifying measures such as stiffness ratio or computational stiffness can be used to decide about this switching. However, eigen-value calculations are computationally extensive and hence are not normally used for large systems of differential equations.

The general approach to solving stiff equations is to use implicit methods. Historically, two chemical engineers, Curtis and Hirschfelder (1952), proposed the first set of numerical formulas that are well suited for stiff initial value problems by adopting

$$y_{n+1} - y_n = h f_{n+1} \tag{3.38}$$

$$y_{n+2} - \frac{4}{3} y_{n+1} + \frac{1}{3} y_n = \frac{2}{3} f_{n+2} \tag{3.39}$$

Both schemes are implicit and belong to the well-known class of backward difference formulas (BDF), the first one being the implicit Euler's scheme.

Consider the same differential equation presented earlier (Equation 3.33) and apply a backward Euler's method, which leads to the following solution:

$$y_n = (1 + 100h)^{-n} + 1 \tag{3.40}$$

Now there is no unstable behavior regarding the size of h. Note that in the explicit Euler's method we were approximating the solution by a polynomial, and there exists no polynomial that can approximate the exponential term as x tends to ∞, hence, the instability. By using the implicit method, we have expressed the solution in the form of a rational function, which can go to zero as x tends to ∞.

A detailed theory of stability and the different definitions of stability criteria (e.g., A-stable systems, stiffly stable systems, etc.) are beyond the scope of this book and readers are advised to look elsewhere for details (Henrici, 1962; Gear, 1971b; Lambert, 1973; Butcher, 1987; Fatunla, 1988). Therefore, we will end this section by presenting below a list of different stiff algorithms available in the literature. This list is taken from Fatunla (1988) and is by no means complete. However, it provides a large number of methods from which to choose.

A List of Stiff Algorithms:

- Backward difference formulas, BDF (Gear, 1969, 1971a, b, 1980; Hindmarsh, 1974, 1980; Byrne and Hindmarsh, 1975; Shampine and Watts, 1979; and Petzold, 1983) and MEBDF (Cash, 1983). Although each method has some advantages and disadvantages, the packages based on BDF methods, like DIFSUB (Gear, 1969), and a variant of LSODE (Hindmarsh, 1980) have been thoroughly tested and are known to be robust and reliable. Also, the LSODA version of LSODE offers an automatic switch from stiff to nonstiff algorithms (Petzold, 1983). Most of the batch distillation codes (Boston et al., 1983; Diwekar and Madhavan, 1991) use computer codes based on BDF.

- Second derivative methods (Enright, 1974a,b; Enright et al., 1975; Sacks-Davis, 1977, 1980; Sacks-Davis and Shampine, 1981; Addison, 1979).

- Extrapolation process (Lindberg, 1972, 1973; Bader and Deuflhard, 1983; Deuflhard, 1983, 1985; and Shampine, 1983a-c).

- Implicit, semi-implicit, and singly implicit Runge–Kutta methods (Butcher, 1964, 1976a, b; Rosenbrock, 1963; Norsett, 1974; Burrage, 1978a, b; Burrage et al., 1980; Wolfbrandt, 1977; Verner, 1979; Varah, 1978; Cash, 1975, 1977a, b; Cash and Singhal, 1982; Bokhoven, 1980; Alexander, 1977; Kaps and Wanner, 1981; Kaps and Rentrop, 1979, 1984; Norsett and Thomsen, 1984, 1986; Verwer, 1981, 1982; Cameron and Gani, 1988).

- Cyclic or composite multistep methods (Bickart and Picel, 1973; Albrecht, 1978, 1985, 1987; Tischer and Sacks-Davis, 1983; Tischer and Gupta, 1985; Isereles, 1984; Zhou, 1986; Rosser, 1967, Chu and Hamilton, 1987).

In general, stiff algorithms can accommodate the use of a large mesh size outside the transient (nonstiff) phase. In the transient phase, automatic codes attempt to identify the optimum mesh size which keeps the local truncation error within the limit of the requested accuracy. Outside the transient (stiff) phase, the growth of propagated errors (instability) controls the choice of mesh size if a nonstiff method is adopted, because the stability restrictions are independent of the accuracy requirements.

Stiff systems can also be defined as near index systems, and index problems as infinitely stiff problems (Chung, 1991). Although this chapter does not deal with index problems, it should be noted that for high-index systems (or very stiff systems), it is difficult to apply any numerical integration method unless the system is transformed in some way to reduce the index of the system. The procedures for transformation tend to be problem dependent and can be very difficult to apply to a large dimensional system. The next few subsections are devoted to the numerical integration of equations governing batch distillation, where the system can become very stiff, approaching index problems, and even stiff integrators may fail to yield the solution. The dynamic model, shown in Table 3.1, is modified to take care of this problem.

3.4   Solution of Column Dynamics

Now that we have seen the numerical integration techniques, we return to the dynamics of batch distillation. The governing equations of batch distillation dynamics have already been discussed in an earlier section. So, we will focus our attention on specific examples.

Example 3.1: A mixture containing 100 moles of a four-component mixture having relative volatilities of 2.0, 1.5, 1.0, and 0.5 is to be distilled in a batch distillation column. The column has eight theoretical plates with a holdup of 2 moles per plate and a condenser holdup of 2 moles. The vapor boilup rate V of the reboiler is 100 moles/hr. Assume an initial total reflux condition and use the steady state model presented in Table 3.2 to obtain the initial values for integration. The column is operating under a constant reflux mode with a reflux ratio equal to 1.0.

a) Simulate a 1 hr operation of the column using a second-order Runge–Kutta method and compare the distillate composition profile with those obtained using the stiff integrator LSODE. Use a step size of 0.01 hr for the Runge–Kutta method.

b) Reduce the plate holdup and condenser holdup from 2 moles to 1 mole/plate. Is the Runge–Kutta method good for simulating the column under this condition?

c) Derive the conditions for variable reflux operation and simulate a 1 hr operation of the column by maintaining the composition of component 1 constant at 0.65 mole fraction.

Solution:

a) Figure 3.2 shows the comparison of distillate composition profiles obtained using the two methods. It can be seen that the second-order Runge–Kutta method performs equally as well as the stiff integrator.

b) For the reduced-holdup column, the results of the two methods are plotted in Figure 3.3. In this case, the two methods show significantly different results. Furthermore, the Runge–Kutta method results also show negative compositions as well as compositions exceeding unity, both of which are physically unfeasible. The conclusion is that the second-order Runge–Kutta method is not appropriate for this problem, owing to the stiff nature of the problem. This suggests the use of stiff integrators like LSODE for the solution of this problem. Figure 3.3 also shows the results for the batch distillation dynamics correctly represented using the LSODE integrator. Most of the batch distillation columns have holdups ranging from 5 to 10 percent and hence need stiff integrators to simulate the column dynamics.

c) In the variable reflux case, the reflux is varied in order to maintain the composition of the key component (component 1 in this case) as constant. While this could be achieved by a feedback controller, it is much easier and more accurate to derive an open-loop control law, as presented below (Diwekar, 1988).

Maintaining $x_D^{(1)}$ constant also implies $dx_D^{(1)}/dt = 0$. Therefore, from Equation 3.12

$$\frac{dx_D^{(1)}}{dt} = \frac{V_1}{H_D}(y_1^{(1)} - x_D^{(1)}) = 0 \qquad (3.41)$$

The objective is to satisfy the above equation by changing R. However, the above equation is not directly related to R. Therefore, differentiation of this equation results in

$$\frac{V_1}{H_D}\left(\frac{dy_1^{(1)}}{dt} - \frac{dx_D^{(1)}}{dt}\right) = 0 \qquad (3.42)$$

By substituting Equation 3.41 in 3.42 and expressing $y_1^{(1)}$ in terms of $x_1^{(1)}$ (as

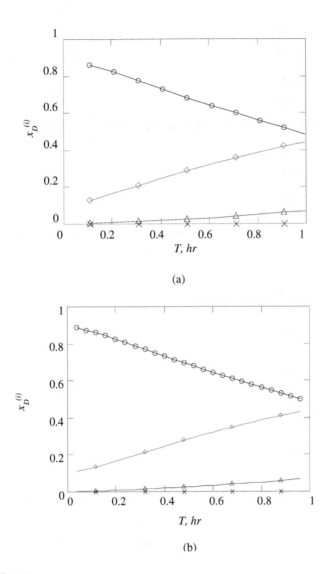

(a)

(b)

FIGURE 3.2
Transient Composition Profiles for Example 3.1a Using (a) a Second-Order
Runge–Kutta Method and (b) the Stiff Integrator LSODE

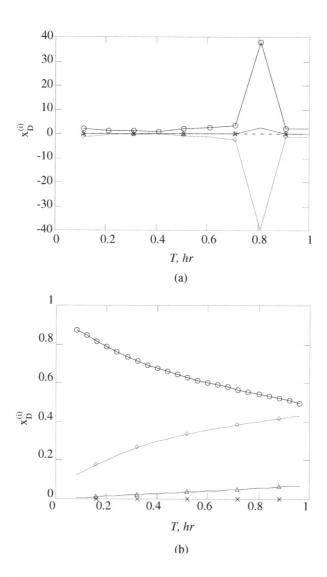

FIGURE 3.3
Transient Composition Profiles for Example 3.1b Using (a) a Second-Order
Runge–Kutta Method and (b) the Stiff Integrator LSODE

$dy_1^{(1)}/dx_1^{(1)} \neq 0)$ we obtain the following equation.

$$\frac{dx_1^{(1)}}{dt} = 0 \qquad (3.43)$$

After simplifying the above equations:

$$R = \frac{1}{\dfrac{(x_D^{(1)} - x_1^{(1)})}{(y_1^{(1)} - y_2^{(1)})} - 1} \qquad (3.44)$$

After the initial reflux condition, the column is operated at zero reflux until it hits the desired distillate composition and then the above equation is used to calculate the variable reflux transient profiles. The results are plotted in Figure 3.4. Diwekar (1988) has observed that the open-loop control law given in Equation 3.44 provides good constant composition control as desired.

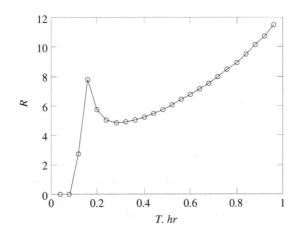

FIGURE 3.4
Transient Composition Profiles for Example 3.1c)

The above example showed how a system changes from nonstiff to stiff if the amount of holdup is changed. The stiffness ratio for the set of batch distillation dynamic equations depends upon a number of factors other than the holdup. Distefano (1968a) provided an approximate measure of the stiffness in terms of physical quantities, such as liquid flowrate L, equilibrium constant K, vapor rate V, and holdup H as the $(L + KV)/H$ ratio. Systems with large volatility differences in components result in a large K, and, hence, a large the $(L + KV)/H$ factor, thereby increasing the stiffness of the system. Similarly, a high reflux results in a high L and a high value of $(L+KV)/H$ factor. Sadamoto and Miyahara (1983) also supported Distefano's approximate analysis.

3.4.1　Batch Distillation Startup

The batch distillation startup period consists of the following three steps:

1. preheating of the still charge to its bubble point,

2. filling the column and the condenser holdups,

3. running without distillate withdrawal and taking the unit to a steady state (Gonzalez-Velasco et al., 1987).

At the end of the total reflux condition (step 3 above), the given charge distributes itself throughout the column. This effect is known as the capacitance effect. Associated with the capacitance effect is the concept of equilibration time, T_{eq}. The equilibration time is the time required to attain the steady state from the time the column is charged. This initial starting operation is usually carried out at a total reflux. The dynamics of the startup condition are described by the equations given in Table 3.3.

The second step can be achieved in different ways, as given below. Depending on the method used, different initial conditions are necessary to solve the differential equations given in Table 3.3:

- Charge the feed from the top so that initially the pot composition as well as the composition on each plate is equal to the feed composition. For this operation, the initial conditions for the dynamic model in Table 3.3 are given by the following equation.

$$x_B^{(i)} \; = \; x_j^{(i)} \; = \; x_F^{(i)}, i = 1, 2, \ldots, n; j = 1, 2, \ldots, N \qquad (3.45)$$

- Charge the feed to the still and operate the column without reflux (zero reflux). This results in the composition on each plate being equal to the vapor composition, which is in equilibrium with the feed composition, leading to the following initial conditions.

$$x_j^{(i)} \; = \; y_F^{(i)}, i = 1, 2, \ldots, n; j = 1, 2, \ldots, N \qquad (3.46)$$

This mode was suggested by Luyben (1971) but has not been applied in simulation or in practice (Gonzalez-Velasco et al., 1987).

Example 3.2: An equimolar mixture containing 100 moles of a four-component mixture having relative volatilities of 2.0, 1.5, 1.0, 0.5 is to be distilled in a batch distillation column. The column has ten theoretical plates with a holdup of 1 mole per plate and a condenser holdup of 1 mole. The vapor boilup rate V of the reboiler is 100 mole/hr. Assume initially a total reflux condition. The column is operating under constant reflux mode with the reflux ratio equal to 1.0. Simulate the operation of the column for 1 hour. Calculate the initial startup conditions:

a) by simulating the complete dynamics of the startup operation given in Table 3.3.

TABLE 3.3
The Startup Operation Dynamic Model

Composition Calculations
Condenser and Accumulator Dynamics
$\frac{dx_D^{(i)}}{dt} = \frac{V_1}{H_D}(y_1^{(i)} - x_D^{(i)}), i = 1, 2, \ldots, n$
Plate Dynamics
$\frac{dx_j^{(i)}}{dt} = \frac{1}{H_j}[V_{j+1}y_{j+1}^{(i)} + L_{j-1}x_{j-1}^{(i)} - V_j y_j^{(i)} - L_j x_j^{(i)}],$
$i = 1, 2, \ldots, n; j = 1, 2, \ldots, N$
Reboiler Dynamics
$\frac{dx_B^{(i)}}{dt} = \frac{1}{B}[L_N(x_N^{(i)} - x_B^{(i)}) - V_B(y_B^{(i)} - x_B^{(i)})],$
$i = 1, 2, \ldots, n$
Flowrate Calculations
At the Top of the Column
$V_1 = L_0$
On the Plates
$L_j = V_{j+1} ; j = 1, 2, \ldots, N$
$V_{j+1} = \frac{1}{J_{j+1} - I_j}[V_j(J_j - I_j) + L_{j-1}(I_j - I_{j-1})]; j = 1, 2, \ldots, N+1$
Heat Duty Calculations
Condenser Duty
$Q_D = V_1(J_1 - I_D) - H_D\delta_t I_D$
Reboiler Duty
$Q_B = V_B(J_B - I_B) - L_N(I_N - I_B) + B\delta_t I_B$
Thermodynamics Correlations
Equilibrium Relations
$y_j^{(i)} = f((x_j^{(k)}, k = 1, \ldots, n), TE_j, P_j)$
Enthalpy Calculations
$I_j = f((x_j^{(k)}, j = 1, \ldots, n), TE_j, P_j)$
$J_j = f((y_j^{(k)}, j = 1, \ldots, n), TE_j, P_j)$

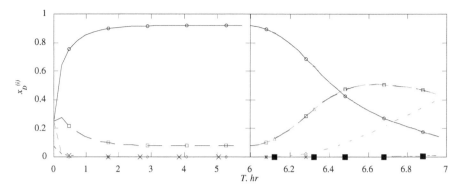

FIGURE 3.5

Transient Composition Profiles for Example 3.2 Using Different Startup Models

   b) using the steady state equations given in Table 3.2. Assume that the equilibration time is approximately 6 hr.

Are the two transient profiles same? Discuss the advantages and limitations of each method.

Solution: Figure 3.5 shows the comparison of distillate composition profiles by the two methods. It can be seen that the simulation profiles obtained from the two methods are identical. Simulating the complete dynamics of the startup operation can be computationally intensive but one can get estimates of equilibration time, which are not possible to obtain from the steady state model. Furthermore, several times in the plant, the total reflux operation is interrupted before the equilibration time is reached. In this situation, one has to resort to complete dynamic simulation of the startup conditions.

3.4.2   Low-Holdup Semirigorous Model

Example 3.3: We have seen that in Example 3.1 the column could be easily simulated using LSODE for all operating conditions. Now, reduce the plate holdup and condenser holdup from 1 mole to 0.01 mole/plate and simulate the operation. Note the results.

Solution: Figure 3.6 shows the transient profiles obtained using LSODE. The LSODE method could not integrate the column because the step size became so small that rounding errors dominated the performance.

The above example shows that for columns where plate dynamics are significantly faster than reboiler dynamics (due to very small plate holdups), stiff integrators often fail to find a solution. The solution to this problem is to split the system into two levels: 1) the reboiler, where the dynamics are slower, can be represented by differential equations, and 2) the rest of the column

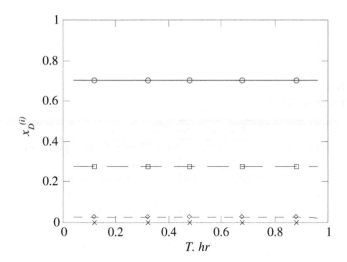

FIGURE 3.6
Transient Composition Profiles for Example 3.3

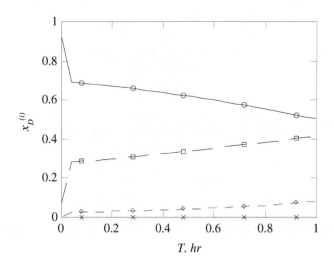

FIGURE 3.7
Transient Composition Profiles for Example 3.3 Using the Semirigorous Model

TABLE 3.4

The Semirigorous Model

Composition Calculations
Condenser and Accumulator Quasi-Steady State Approximation
$0 = (y_1^{(i)} - x_D^{(i)}), i = 1, 2, \ldots, n$
Plates, Quasi-Steady State Approximation
$0 = [V_{j+1}y_{j+1}^{(i)} + L_{j-1}x_{j-1}^{(i)} - V_j y_j^{(i)} - L_j x_j^{(i)}],$
$i = 1, 2, \ldots, n; j = 1, 2, \ldots, N$
Reboiler Dynamics
$\frac{dx_B^{(i)}}{dt} = \frac{1}{B}[L_N(x_N^{(i)} - x_B^{(i)}) - V_B(y_B^{(i)} - x_B^{(i)})], \quad x_{B0}^{(i)} = x_F^{(i)}$
$, i = 1, 2, \ldots, n$
Flowrate Calculations
At the Top of the Column
$L_0 = R\frac{dD}{dt}$
$V_1 = (R+1)\frac{dD}{dt}$
On the Plates
$L_j = V_{j+1} + L_{j-1} - V_j; j = 1, 2, \ldots, N$
$V_{j+1} = \frac{1}{J_{j+1} - I_j}[V_j(J_j - I_j) + L_{j-1}(I_{j-1} - I_j]j = 1, 2, \ldots, N+1$
At the Bottom of the Column
$\frac{dB}{dt} = L_N - V_B; \quad B_0 = F$
Heat Duty Calculations
Condenser Duty
$Q_D = V_1(J_1 - I_D)$
Reboiler Duty
$Q_B = V_B(J_B - I_B) - L_N(I_N - I_B) + B\delta_t I_B$
Thermodynamic Equilibrium and Enthalpy Relations
Equilibrium Relations
$y_j^{(i)} = f((x_j^{(k)}, k = 1, \ldots, n), TE_j, P_j)$
Enthalpy Calculations
$I_j = f((x_j^{(k)}, j = 1, \ldots, n), TE_j, P_j)$
$J_j = f((y_j^{(k)}, j = 1, \ldots, n), TE_j, P_j)$

can be assumed to be in the quasi-steady state. This results in a zero holdup model, which is given in Table 3.4. This approach is used for a semirigorous simulation of batch distillation operation in the software package BATCH-DIST (Diwekar, 1988; Diwekar and Madhavan, 1991). Figure 3.7 shows the results of this model using the fourth-order Runge–Kutta method. The model results are closer to those obtained previously in Example 3.1. The holdup effects can be neglected in a number of cases where this model approximates the column behavior accurately. However, note that this model involves an iterative solution of nonlinear plate-to-plate algebraic equations, which can be computationally less efficient than the dynamic model. This low-holdup semirigorous model is used in many studies such as Domench and Enjalbert (1981), Doherty and co-workers (Bernot et al., 1990, 1991), and Farhat et al. (1990).

References

Addison C. A. (1979), Implementing a stiff method based upon the second derivative formula, *Technical Report No. 130*, Department of Computer Science, University of Toronto, Canada.

Albrecht P. (1978), On the order of composite multistep methods for ordinary differential equations, *Numerische Mathematik*, **29**, 381.

Albrecht P. (1985), Numerical treatment of ODEs: The theory of A-methods, *Numerische Mathematik*, **47**, 59.

Albrecht P. (1987), A new theoretical approach to R-K methods, *SIAM Journal of Numerical Analysis*, **24**, 391.

Alexander R. (1977), Diagonally implicit R-K methods for stiff ordinary differential equations, *SIAM Journal of Numerical Analysis*, **14**, 1006.

Bader G. and P. Deuflhard (1983), A semi-implicit midpoint rule for stiff systems of ODEs, *Numerische Mathematik*, **41**, 373.

Bernot C., M. F. Doherty, and M. F. Malone (1990), Patterns of composition changes in multicomponent batch distillation, *Chemical Engineering Science*, **45**, 1207.

Bernot C., M. F. Doherty, and M. F. Malone (1991), Feasibility and separation sequencing in multicomponent batch distillation, *Chemical Engineering Science*, **46**, 1311.

Bickart T. A. and Z. Picel (1973), High order stiffly stable composite multistep methods for the numerical integration of stiff differential equations, *BIT*, **13**, 272.

Bokhoven V. (1980), Efficient high order one-step method for integration of stiff differential equations, *BIT*, **20**, 34.

Boston J. F., H. I. Britt, S. Jiraponghan, and V. B. Shah (1983), An advanced system for the simulation of batch distillation operation, *Foundations of Computer Aided Process Design*, Mah R. S.H. and W. Seider Eds., **2**, 203.

Burrage K. (1978a), A special family of Runge–Kutta methods, *BIT*, **18**, 22.

Burrage K. (1978b), High order algebraically stable R-K methods, *BIT*, **18**, 373.

Burrage K., J. C. Butcher, and F. H. Chipman (1980), An implementation of singly implicit R-K methods, *BIT*, **20**, 326.

Butcher J. C. (1964), Implicit Runge–Kutta processes, *Mathematics of Computation*, **18**, 50.

Butcher J. C. (1976a), On the implementation of implicit R-K methods, *BIT*, **16**, 237.

Butcher J. C. (1976b), Implicit R-K and related methods, In *Modern Numerical Methods for Ordinary Differential Equations* (G. Hall and J. M. Watts, eds.), Oxford University Press, Oxford, 136-151.

Butcher J. C. (1987), *The Numerical Analysis of Ordinary Differential Equations*, John Wiley & Sons, NY.

Byrne G. D. and A. C. Hindmarsh (1975), EPISODE: An experimental package for integration of systems of ordinary differential equations, Report UCID-30112, Lawrence Livermore Laboratory, Livermore,CA.

Cameron I. T. and R. Gani (1988), Adaptive Runge–Kutta algorithms for dynamic simulation, *Computers and Chemical Engineering*, **12**, 705.

Cameron I. T., C. A. Ruiz, and R. Gani (1986), A generalized model for distillation columns-II: Numerical and computational aspects, *Computers and Chemical Engineering*, **10**, 199.

Cash J. R. (1975), A class of implicit R-K methods for the numerical integration of ODEs, *Journal of the ACM*, **22**, 503.

Cash J. R. (1977a), Semi-implicit R-K procedures with error estimates for the numerical solution of stiff systems of ODEs, *Journal of the ACM*, **24**, 455.

Cash J. R. (1977b), On a class of implicit RK procedures, *Journal of Institute of Mathematics and Its Applications*, **19**, 455.

Cash J. R. (1977c), A note on the computational aspects of a class of IRK procedures, *Journal of Institute of Mathematics and Its Applications*, **20**, 425.

Cash J.R. (1983), The integration of stiff IVPs in ordinary differential equations using a modified extended BDF, *Computers and Mathematics with Applications*, **9**, 645.

Cash J. R. and A. Singhal (1982), Mono-implicit RKF for the numerical integration of stiff differential equations, *IMA Journal on Numerical Analysis*, **2**, 211.

Chu M. T. and H. Hamilton (1987), Parallel solution of ordinary differential equations by multiblock methods, *SIAM Journal of Scientific and Statistical Computing*, **8**, 342.

Chung Y. (1991), *Solving index and near index problems in dynamic simulation*, Ph.D. Thesis, Department of Chemical Engineering, Carnegie Mellon University, Pittsburgh, PA.

Cullie P. E. and G. V. Reklaitis (1986), Dynamic simulation of multicomponent batch rectification with chemical reactions, *Computers and Chemical Engineering*, **10**, 389.

Curtis C. F. and J. O. Hirschfelder (1952), Integration of stiff equations, *Proceedings of National Academy of Sciences*, **38**, 235.

Deuflhard P. (1983), Order and stepsize control in extrapolation method, *Numerische Mathematik*, **41**, 399.

Deuflhard P. (1985), Recent progress in extrapolation method for ODEs, *SIAM Review*, **27**, 505.

Distefano G. P. (1968a), Mathematical modeling and numerical integration of multicomponent batch distillation equations, *AIChE Journal*, **14**, 190.

Distefano G. P. (1968b), Stability of numerical integration techniques, *AIChE Journal*, **14**, 946.

Diwekar U. M. (1988), *Simulation, design, and optimization of multicomponent batch distillation columns*, Ph.D. thesis, Indian Institute of Technology, Bombay, India.

Diwekar U. M. and K. P. Madhavan (1991), BATCH-DIST: A comprehensive package for simulation, design, optimization, and optimal control of multicomponent, multifraction batch distillation columns, *Computers and Chemical Engineering*, **15**, 833.

Domench S. and N. Enjalbert (1981), Program for simulating batch rectification as a unit operation, *Computers and Chemical Engineering*, **5**, 181.

Enright W. H. (1974a), Second derivative multistep methods for stiff ODEs, *SIAM Journal of Numerical Analysis*, **11**, 321.

Enright W. H. (1974b), Optimal second derivative methods for stiff systems, *Conference on Stiff Differential Systems* (R. A. Wiloughby, ed.), Plenum Publishing Co., New York, pp. 95-109.

Enright W. H., T. E. Hull, and B. Lindberg (1975), Comparing numerical methods for stiff systems of ODEs, *BIT*, **15**, 10.

Farhat S., M. Czernicki, L. Pibouleau, and S. Domench (1990), Optimization of multiple-fraction batch distillation by nonlinear programming, *AIChE Journal*, **36**, 1349.

Fatunla S. O. (1988), *Numerical Methods for Initial Value Problems in Ordinary Differential Equations*, Academic Press, San Diego, CA.

Finlayson B. A. (1980), *Nonlinear Analysis in Chemical Engineering*, McGraw-Hill, NY.

Gear C. W. (1969), The automatic integration of stiff ODEs, in *Information Processing*, **68** (A.J.H. Morrell, ed.), North-Holland Publishing Company, Amsterdam, 187.

Gear C. W. (1971a), Algorithm 407: DIFSUB for solution of ODEs, *Communications of ACM*, **14**, 185.

Gear C. W. (1980), Unified modified divided implementation of ADAMS and BDF formulas, Report No. UIUCDCS-R-80-1014., Department of Computer Science, University of Illinois at Urbana-Champaign, IL.

Gear C. W. (1971b), *Numerical Initial Value Problems in Ordinary Differential Equations*, Prentice-Hall Inc., Englewood Cliffs, NJ.

Golub G. H. and J. M. Ortega (1992), *Scientific Computing and Differential Equations*, Academic Press, San Diego, CA.

Gonzalez-Velasco J. R., M. A. Gutierrez-Ortiz, J. S. Catresana-Pelayo, and J. A. Gonzalez-Marcos (1987). Improvements in batch distillation startup, *Industrial and Engineering Chemistry Research*, **26**, 745.

Hairer E., S. Norsett, and G. Wanner (1987), *Solving Ordinary Differential Equations; I. Nonstiff problems*, Springer-Verlag, NY.

Henrici P. (1962), *Discrete Variable Methods in ODEs*, John Wiley & Sons, New York.

Hindmarsh A. C. (1974), GEAR: ODE system solver, Revision 3, Report No. UCID-30001, Lawrence Livermore Laboratory, University of California, Livermore, CA.

Hindmarsh A. C. (1980), LSODE and LSODI: Two new initial value ODE solvers, *ACM SIGNUM Newsletter*, **15**, 10.

Huckaba C. E. and D. E. Danly (1960), Calculation procedures for binary batch rectification, *AIChE Journal*, **6**, 335.

Isereles A. (1984), Composite methods for numerical solution of stiff ODEs, *SIAM Journal of Numerical Analysis*, **21**, 340.

Kaps P. and P. Rentrop (1979), Generalized R-K methods of order four with stepsize control for stiff ODEs, *Numerische Mathematik*, **33**, 55.

Kaps P. and P. Rentrop (1984), Application of variable order semi-implicit RK method to chemical models, *Computers and Chemical Engineering*, **8**, 393.

Kaps P. and G. Wanner (1981), A study of Rosenbrock-type methods for stiff ODEs: A comparison of Richardson extrapolation and the embedding technique, *Numerische Mathematik*, **38**, 279.

Lambert J. (1973), *Computational Methods in Ordinary Differential Equations*, Wiley & Sons, NY.

Lindberg B. (1972), IMPEX–A program package for solution of systems of stiff differential equations, Report No. NA:7250, Department of Information Processing, Royal Institute of Technology, Stockholm, Sweden.

Lindberg B. (1973), IMPEX 2– A procedure for solution of systems of stiff differential equations, Report No. TRITA-NA:7303, Department of Information Processing, Royal Institute of Technology, Stockholm, Sweden.

Luyben W. L. (1971), Some practical aspects of optimal batch distillation design, *Industrial and Engineering Chemistry Process Design, Development*, **10**, 54.

Meadows E. L. (1963), Multicomponent batch distillation calculations on a digital computer, *Chemical Engineering Symposium Series 46*, **59**, 48.

Norsett S. P. (1974), Semi-explicit R-K Methods, Report No. **6/74b**, Department of Mathematics, University of Trondheim, Norway.

Norsett S. P. and P. G. Thomsen (1984), Embedded SDIRK methods of basic order three, *BIT*, **24**, 634.

Norsett S. P. and P. G. Thomsen (1986), SIMPLE, a stiff system solver, ODE conference held at Sandia National Laboratory, Albuquerque, NM, July, 1986.

Petzold L. (1983), Automatic selection of method for solving stiff and nonstiff systems of ODEs, *SIAM Journal of Scientific Computing*, **4**, 136.

Rosenbrock H. H. (1963), Some general implicit processes for numerical solution of differential equations, *Computer Journal*, **5**, 329.

Rosser J. B. (1967), A R-K for all seasons, *SIAM Review*, **9**, 417.

Sacks-Davis R. (1977), Solution of stiff ODEs by a second derivative method, *SIAM Journal of Numerical Analysis*, **14**, 1088.

Sacks-Davis R. (1980), Fixed leading coefficient implementation of second derivative formulas for stiff ODEs, *ACM Transactions on Mathematical Software*, **6**, 540.

Sacks-Davis R. and L. F. Shampine (1981), A type insensitive ODE code based on second derivative formula, *Computations and Mathematics with Applications*, **7**, 487.

Sadamoto H. and K. Miyahara (1983), Calculation procedure for multicomponent batch distillation, *International Chemical Engineering*, **23**, 56.

Shampine L. F. (1983a), Efficient extrapolation methods for ODEs I, *IMA Journal of Numeral Analysis*, **3**, 383.

Shampine L. F. (1983b), Type sensitive ODE codes based on extrapolation, *SIAM Journal of Scientific and Statistical Computing*, **4**, 635.

Shampine L. F. (1983c), Efficient extrapolation method for ODEs II, Report No. SAND-83-1927, Sandia National Laboratory, Albuquerque, NM.

Shampine L. F. and H. A. Watts (1979), DEPAC–Design of a user oriented package of ODE solvers, Report No. SAND79-2374, Sandia National Laboratory, Albuquerque, NM.

Tischer P. E. and G. K. Gupta (1985), An evaluation of some new cyclic multistep formulas for stiff ordinary differential equations, *ACM Transactions on Mathematical Software*, **11**, 263.

Tischer P. E. and R. Sacks-Davis (1983), A new class of cyclic multistep formulae for stiff systems, *SIAM Journal of Scientific and Statistical Computing*, **4**, 733.

Varah J. M. (1979), On the efficient implementation of implicit R-K methods, *Mathematics of Computation*, **33**, 557.

Verner J.H. (1978), Explicit R-K methods with estimates of the local truncation error, *SIAM Journal of Numerical Analysis*, **15**, 772.

Verwer J. G. (1981), An analysis of Rosenbrock methods for nonlinear stiff IVPs, *SIAM Journal of Numerical Analysis*, **19**, 155.

Verwer J. G. (1982), Instructive experiments with some R-K Rosenbrock methods, *Computation and Mathematics with Applications*, **8**, 217.

Wolfbrandt A. (1977), A study of Rosenbrock processes with respect to order condition and stiff stability, Department of Computer Science, Chalmers University of Technology and University of Goteborg, Sweden.

Zhou B. (1986), A-stable and L-stable block implicit one-step methods, ODE conference held at Sandia National Laboratory, Albuquerque, NM, July, 1986.

Exercises

3.1 An equimolar mixture containing 100 moles of a three-component mixture having relative volatilities of 1.7, 1.16, 1.0 and is to be distilled in a batch distillation column. The column has ten theoretical plates with a holdup of 2 moles per plate and a condenser holdup of 2 moles. The vapor boilup rate V of the reboiler is 100 mole/hr. Assume an initial total reflux condition and use the steady state model presented in Table 3.2 to obtain the initial values for integration. The column is operating under a constant reflux mode with a reflux ratio equal to 2.0.

    a) Simulate a 1 hr operation of the column using a second-order Runge–Kutta method and compare the distillate composition profile with those obtained using the stiff integrator LSODE. Use a step size of 0.01 hr for the Runge–Kutta method.

    b) Reduce the plate holdup and condenser holdup from 2 moles to 0.5 mole/plate. Is the Runge–Kutta method good for simulating the column under this condition?

    c) Simulate a 1 hr operation of the column by maintaining composition of component 1 constant at 0.65 mole fraction.

3.2 Simulate the transient total reflux condition for the column given in the above example. What is the equilibration time?

3.3 Change the plate holdup for the above column slowly and find out the switching point for the use of the semirigorous model.

3.4 Show that the approximate measure of stiffness for a batch distillation column can be given by $(L + KV)/H$.

3.5 Define an index problem and describe its relation to stiff systems.

4

SIMPLIFIED MODELS

CONTENTS

Chapter 3 described the rigorous model for batch distillation, as well as the difficulties in solving the dynamic equations constituting the rigorous models. This chapter is devoted to simplified models of batch distillation. Since computers have made it possible to analyze and solve large-scale models in a reasonable amount of time, what is the need for developing simplified models? The first section deals with this question. It is followed by the development of a simple, efficient, reasonably accurate shortcut model. The chapter also describes a hierarchy of models, ranging from the simplified shortcut model to the model based on the collocation approach. This chapter also explains the assumptions and the level of abstraction behind each model.

4.1 Need for Simplification

As seen in the last chapter, a rigorous modeling of batch distillation operation involves solution of several stiff differential equations. The computational intensity and memory requirement of the problem increases with increasing the number of plates and components. For an n-component system and a column with N plates, the simulation of constant reflux or variable reflux operation requires a simultaneous solution of $n \times N$ material balance differential equations, N energy balance differential or difference equations, plus the vapor-liquid equilibrium calculation associated with each differential material balance equation. The vapor-liquid equilibria calculations are iterative in nature and become more complicated for nonideal systems. The computational complexity associated with rigorous models does not allow us to derive global properties, such as feasible regions of operation, which are critical for optimization, optimal control, and synthesis problems. Even if such information were available, the computational costs of optimization, optimal control, or synthesis using rigorous models are prohibitive.

One way to deal with the problems associated with rigorous models is to develop simplified models. Chapter 2 described one of the early models of batch distillation which was limited to graphical integration applied to batch distillation using the Rayleigh equation for binary systems. The empirical shortcut models for binary systems succeeded the graphical models. These empirical models had limited applicability and, hence, were only described in the context of two examples in the exercise section of Chapter 2. These models were based on the concept of a Standard Separation Curve (Rose and Welshans, 1940) and the idea of a Pole Height proposed by Zuiderweg (1953).

Development of simplified models for batch distillation and their use in optimization and optimal control seems to be the recent trend in batch distillation studies (Farhat et al., 1990; Chiotti and Iribarren, 1991; Al-Tuwaim and Luyben, 1991). These models are either confined to binary (Chiotti and Iribarren, 1991) or ternary systems (Al-Tuwaim and Luyben, 1991), or they use the semirigorous plate-to-plate calculation method for the simulation (Farhat et al., 1990).

BATCH-DIST, a comprehensive package for multicomponent, multifraction batch distillation columns (Diwekar and Madhavan, 1991b), provides a hierarchy of models for simulation, ranging from the most simplified model to the most rigorous model based on Distefano's analysis (described in Chapter 3). The most simplified model in BATCH-DIST uses the shortcut method (Diwekar, 1988; Diwekar and Madhavan, 1991a) and is applicable to nearly ideal systems and columns where the holdup effects are insignificant. The model rigor, data requirement, and computational effort increase as one climbs higher in the hierarchy. These simplified models are abstractions of rigorous models, and their accuracy depends on the simplifying assumptions embedded within them. The process of abstraction can be viewed as a trade-off between simplicity and accuracy. The usefulness of abstracted models depends on the ease with which they can be analyzed for global behaviors without compromising accuracy. Moreover, abstracted models are expected to be computationally simpler to analyze.

The schematic of the process of abstraction for batch distillation column models is shown in Figure 4.1. The abstract (simplified) model is derived from the rigorous model through a set of simplifying assumptions. The abstract model in Figure 4.1 is the shortcut method, which essentially reduces the number of equations from $n \times N$ material balance differential equations, N energy balance differential or difference equations, and the vapor-liquid equilibrium calculation associated with each differential material balance equation $(2(n \times N) + N)$ to *two* differential equations describing the dynamics of the reboiler and $2n$ algebraic equations for the top and bottom compositions and *one* equation relating the design variables, such as reflux ratio and number of plates to these compositions $(2n + 3)$. The details of this method are described in section 4.2, followed by a discussion of the next model in the hierarchy, which takes into account the effect of holdup, especially in the context of the total reflux initial operation.

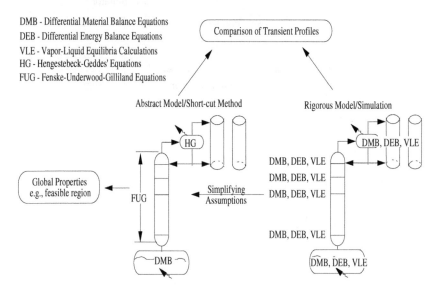

FIGURE 4.1
Abstraction Process

4.2 Shortcut Method

The shortcut method for batch distillation proposed by Diwekar and co-workers (Diwekar, 1988; Diwekar and Madhavan, 1991a; Diwekar, 1991) is based on the assumption that the batch distillation column can be considered as a continuous distillation column with changing feed at any instant (Figure 4.2). Since continuous distillation theory is well developed and tested, the shortcut method (Fenske-Underwood-Gilliland method) for continuous distillation (FUG) is modified for batch distillation. The other assumptions of the shortcut method include constant molal overflow and negligible plate holdups. The details of this method are described below.

For the system shown in Figure 4.2 we assume that the distillation is carried out at constant boilup rate V. The constant molal overflow assumption leads to the following overall material balance equation (refer to Equation 2.7 from Chapter 2).

$$\frac{dB}{dt} = -\frac{V}{R+1}, \quad B_0 = F \qquad (4.1)$$

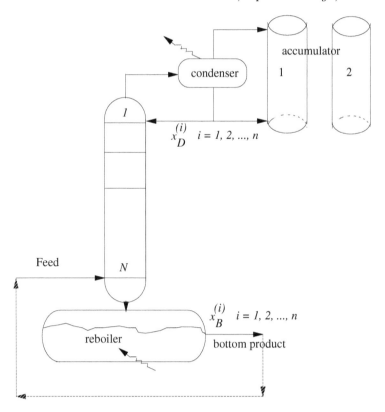

FIGURE 4.2
The Shortcut Method

where F is the feed. A material balance for the key component k over the differential time can be written as:

$$x_D^{(k)} \frac{V}{R+1} = - \frac{d(x_B^{(k)} B)}{dt} \tag{4.2}$$

Substitution of Equation 4.1 in Equation 4.2 results in

$$\frac{dx_B^{(k)}}{dt} = \frac{V}{(R+1)B}(x_B^{(k)} - x_D^{(k)}), \qquad x_{B_0}^{(k)} = x_F^{(k)} \tag{4.3}$$

Since holdup on the plates is ignored in the shortcut model, the Rayleigh equation (Equation 2.1) described in Chapter 2 is applicable here. This leads to the time-implicit equations relating the variation of composition of component i in an n component mixture to the variation of the bottom composition of component k as given below.

$$\frac{dx_B^{(i)}}{(x_D^{(i)} - x_B^{(i)})} = \frac{dx_B^{(k)}}{(x_D^{(k)} - x_B^{(k)})}, \qquad x_{B0}^{(i)} = x_F^{(i)}; i = 1, 2, i \neq k, \ldots, n \quad (4.4)$$

The finite-difference approximation of the above equations results in

$$x_{B_{new}}^{(i)} = x_{B_{old}}^{(i)} + \frac{\triangle x_B^{(k)}\left(x_D^{(i)} - x_B^{(i)}\right)_{old}}{\left(x_D^{(k)} - x_B^{(k)}\right)_{old}}, \qquad i = 1, 2, \ldots, n \quad (4.5)$$

As seen earlier in Chapter 2, the functional relationship between the distillate composition x_D and x_B is crucial for the simulation of the complete operation. It is here that we use the design equations described by the modified FUG method for batch distillation, as discussed below.

Functional Relationship Between x_D and x_B:

At each instant, there is a change in the still composition of the key component, resulting in changes in the still composition of all the other components calculated by the differential material balance equations (Equation 4.5). The Hengestebeck–Geddes equation, given below, relates the distillate compositions to the new bottom compositions in terms of a constant C_1.

The Hengestebeck–Geddes equation

$$x_D^{(i)} = \left(\frac{\alpha_i}{\alpha_1}\right)^{C_1} \frac{x_D^{(k)}}{x_B^{(k)}} x_B^{(i)}, i = 1, 2, \ldots, n (i \neq k) \quad (4.6)$$

It can be easily proved that the constant C_1 of the Hengestebeck–Geddes equation is equivalent to the minimum number of plates, N_{min}, in the Fenske equation. The minimum number of plates is defined as the number of equilibrium plates required for a given separation operating under a total reflux condition (infinite reflux ratio). The minimum number of plates and the minimum reflux described later in this section define the boundary of the operating conditions. Fenske (1932) derived an equation for calculating the minimum number of plates for the constant relative volatility mixtures in terms of the bottom and distillate compositions for continuous columns. The minimum number of plates can be obtained using the Fenske equation, which is normally in terms of two components, say i and k, as given below.

The Fenske equation:

$$N_{min} = \frac{\ln\left[\frac{x_D^{(i)}}{x_D^{(k)}} \frac{x_B^{(k)}}{x_B^{(i)}}\right]}{\ln[\alpha_i]} \quad (4.7)$$

From Equations 4.6 and 4.7, it can be seen that:

$$C_1 \equiv N_{min} \quad (4.8)$$

For a binary mixture, the separation is between only two components; in the case of a multicomponent mixture, however, the separation can be expressed in terms of a binary mixture of two key components, LK, the lightest component appearing in the bottom, and HK, the heavy key component defined as the heaviest component in the top. Unlike continuous distillation, the key components keep changing for batch distillation as time progresses. For example, the light key component will become lighter than light key, and the next component in the relative volatility hierarchy will become the light key component.

At minimum reflux (R_{min}), an infinite number of equilibrium stages are required to achieve the given separation. In the case of continuous distillation, Underwood (1948) developed equations to calculate R_{min}. These equations are given below.

The Underwood equations:

$$\sum_{i=1}^{n} \frac{\alpha_i x_F^{(i)}}{\alpha_i - \phi} = 1 - q \qquad (4.9)$$

$$R_{min} + 1 = \sum_{i=1}^{n} \frac{\alpha_i x_D^{(i)}}{\alpha_i - \phi} \qquad (4.10)$$

The q in the above equations represents the feed condition and is defined as the ratio of heat required to vaporize 1 mol of the feed to the molar latent heat of the feed. The ϕ is the root of the Underwood equation which lies between the α_{LK} and α_{HK}. The shortcut method assumes that batch distillation can be considered as continuous distillation with changing feed. In other words, the bottom product of one time step forms the feed for the next time step. This is equivalent to having the bottom plate as the feed plate and the feed at its boiling point, which means q is unity. Also, the feed composition in Underwood equations can be replaced by the bottom composition. Therefore, Underwood equations for the batch distillation column shown in Figure 4.2 are given by

The Underwood equations

$$\sum_{i=1}^{n} \frac{\alpha_i x_B^{(i)}}{\alpha_i - \phi} = 0 \qquad (4.11)$$

$$R_{min} + 1 = \sum_{i=1}^{n} \frac{\alpha_i x_D^{(i)}}{\alpha_i - \phi} \qquad (4.12)$$

Design variables of the column such as the reflux ratio R and the number of plates N are related to each other by Gilliland's correlation (Gilliland, 1940) through the values of R_{min} and N_{min}.

Gilliland's correlation

$$Y = 1 - \exp\left[\frac{(1 + 54.4X)(X - 1)}{(11 + 117.2X)\sqrt{X}}\right] \qquad (4.13)$$

in which

$$X = \frac{R - R_{min}}{R + 1}; \qquad Y = \frac{N - N_{min}}{N + 1}$$

From the above equations, it can be seen that the shortcut method has only two differential equations (Equations 4.1 and 4.3) and the rest of the equations are algebraic. Because of the algebraic equation oriented form of the shortcut method, it is possible to adapt the model for optimal control calculations very easily. The detailed development of the shortcut method for different operating modes is available in the literature from 1987 (Diwekar et al., 1987; Diwekar, 1988; Diwekar and Madhavan, 1991a), and a unified shortcut method approach to the different operating modes was later presented in Diwekar (1991) and Diwekar (1992a). Table 4.1 shows the time-implicit model for the three operating modes. It can be seen that the same equations in different forms can be used to simulate the different operating modes, namely, variable reflux, constant reflux, and optimal reflux.

At any instant of time Equations 4.1, 4.3, and 4.5 can be used to calculate the bottom composition of all the components. The following procedure is then used to calculate the distillate composition at that instant. The procedure is repeated at each time step until the stopping criterion is satisfied. The complete simulation of the shortcut method for the different operating modes is illustrated in the flowchart shown in Figure 4.3.

The constant C_1 in the Hengestebeck–Geddes equation is equivalent to the minimum number of plates, N_{min}, in the Fenske equation. At this stage, the variable reflux operating mode has C_1 and R, the constant reflux has $x_D^{(k)}$ and C_1, and the optimal reflux has $x_D^{(k)}$, C_1, and R as unknowns. Summation of distillate compositions can be used to obtain C_1 for variable reflux and $x_D^{(k)}$ for both constant reflux and optimal reflux operations, and the FUG equations to obtain R for variable reflux and C_1 for both constant reflux and optimal reflux operations. The optimal reflux mode of operation has an additional unknown, R, which is calculated using the concept of optimizing the Hamiltonian, formulated using the different optimal control methods. The details of the shortcut method for the optimal reflux operation are presented in Chapter 8. However, the method is described qualitatively here for the sake of completeness of the unified approach.

The following procedure illustrates the shortcut method for a single time step in detail for the constant reflux and the variable reflux modes of operation. Note that choosing the proper light key and heavy key components is crucial for the success of this method. Example 4.2 illustrates how it is possible to obtain totally different solutions if one does not choose the correct LK and HK initially.

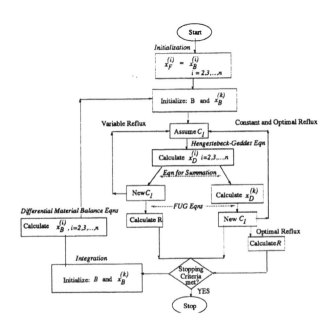

FIGURE 4.3
Flowchart for the Shortcut Method

TABLE 4.1
Time-Implicit Model Equations for the Shortcut Method

Variable Reflux	Constant Reflux	Optimal Reflux
Differential Material Balance Equation		
$x_{B_{new}}^{(i)} = x_{B_{old}}^{(i)} + \dfrac{\Delta x_B^{(k)}\left(x_D^{(i)} - x_B^{(i)}\right)_{old}}{\left(x_D^{(k)} - x_B^{(k)}\right)_{old}}, i = 1, 2, \ldots, n$		
Hengestebeck–Geddes Equation		
$x_D^{(i)} = \left(\dfrac{\alpha_i}{\alpha_k}\right)^{C_1} \dfrac{x_D^{(k)}}{x_B^{(k)}} x_B^{(i)}, i = 2, 3, \ldots, n$		
Unknowns		
R, C_1	$C_1, x_D^{(k)}$	$R, C_1, x_D^{(k)}$
Summation of Fractions		
	$\sum_{i=1}^{n} x_D^{(i)} = 1$	
C_1 Estimation	*$x_D^{(k)}$ estimation*	
$\sum_{i=1}^{n}\left(\dfrac{\alpha_i}{\alpha_k}\right)^{C_1}\dfrac{x_D^{(k)}}{x_B^{(k)}}x_B^{(i)} = 1$	$x_D^{(k)} = \dfrac{1}{\sum_{i=1}^{n}\left(\frac{\alpha_i}{\alpha_k}\right)^{C_1}\frac{x_B^{(i)}}{x_B^{(k)}}}$	
Fenske Equation		
	$N_{min} \approx C_1$	
Underwood Equations		
	$\sum_{i=1}^{n}\dfrac{\alpha_i x_B^{(i)}}{\alpha_i - \phi} = 0; \quad R_{min\ u} + 1 = \sum_{i=1}^{n}\dfrac{\alpha_i x_D^{(i)}}{\alpha_i - \phi}$	
Gilliland's Correlation		
R Estimation	*C_1 Estimation*	
$R = F(N, N_{min}, R_{min\ u})$	$R_{min\ g} = F(N, N_{min}, R)$	
	$\dfrac{R_{min\ g}}{R} - \dfrac{R_{min\ u}}{R} = 0$	
		R Estimation
		$R = F(Minimum\ H)$

Constant Reflux: Given the bottom composition at each time step, the distillate composition is given by the Hengestebeck–Geddes equation (Equation 4.6). However, there are two unknowns, $x_D^{(k)}$ and C_1, in the equation.

1. Assume the initial value of C_1.

2. The Hengestebeck–Geddes equation and the fact that the summation of all the distillate compositions should be equal to unity lead to:

$$\sum_{i=1}^{n}\left(\frac{\alpha_i}{\alpha_k}\right)^{C_1}\frac{x_D^{(k)}}{x_B^{(k)}}x_B^{(i)} = 1 \qquad (4.14)$$

So, the key distillate composition $x_D^{(k)}$ can be obtained as:

$$x_D^{(k)} = 1/\sum_{i=1}^{n} \left(\frac{\alpha_i}{\alpha_k}\right)^{C_1} \frac{x_B^{(i)}}{x_B^{(k)}} \qquad (4.15)$$

Find the value of $x_D^{(k)}$ using the above equation.

3. Substitute the value in the Hengestebeck–Geddes equation and find the distillate composition, $x_D^{(i)}$, $i = 1, 2, \ldots, n$.

4. Solve the Underwood equations by substituting the values of the distillate and the bottom composition for R_{min}. Let us denote this value as $R_{min\ u}$ derived from the Underwod equations.

5. Given that C_1 is equivalent to N_{min} and at a constant reflux R, the reflux ratio is known, and N the number of plates (theoretical) is specified, calculate the value of R_{min} from Gilliland's correlation. Following the same argument used in the case of the Underwood equations this value can be represented as $R_{min\ g}$.

6. For the right value of C_1 both the minimum reflux values $R_{min\ u}$ and $R_{min\ g}$ should be equal. This provides the following equation, where the value of G_c should go to zero when C_1 is converged.

$$G_c = \frac{R_{min\ u}}{R} - \frac{R_{min\ g}}{R} \qquad (4.16)$$

7. Calculate G_c and find whether it is zero within a tolerance. If G_c is approximately zero, the solution is converged for this time step; otherwise use the Newton-Raphson method to calculate the new C_1, and repeat all the steps from step 2.

Example 4.1: A ternary mixture of meta-, ortho-, para-mono-nitro-toluene is to be distilled using a batch distillation column with ten theoretical plates. The column is fed with 100 moles of the mixture with the feed composition 0.6, 0.04, and 0.36 of meta-, ortho-, para-mono-nitro-toluene, respectively. Assume relative volatilities α_{m-p} and α_{o-p} are 1.7 and 1.16, respectively.

a) Simulate the first time step operation, if the column is operating in a constant reflux mode with a reflux ratio equal to 1.0. Assume $k = 1$, $LK = 1$, and $HK = 3$. Assume the initial value of C_1 equal to 2.3691.

b) Simulate a 1 hr operation of the column if the vapor boilup rate V is 100 mole/hr and compare with the results of the simulation using the rigorous model presented in Chapter 3.

Solution:

a) Assuming that initially $C_1 = 2.3691$, from the Hengestebeck–Geddes equation the value of $x_D^{(1)}$ is:

$$x_D^{(1)} = 1/\left(\frac{1.7}{1.7}\right)^{2.3691}\frac{0.6}{0.6} + \left(\frac{1.16}{1.7}\right)^{2.3691}\frac{0.04}{0.6}$$

$$+ \left(\frac{1.0}{1.17}\right)^{2.3691}\frac{0.36}{0.6}$$

$$= 0.8350$$

Therefore, $x_D^{(i)}$, $i = 2, 3$ are given by:

$$x_D^{(2)} = \left(\frac{1.16}{1.7}\right)^{2.3691}\frac{0.8350}{0.6}0.04 = 0.0225$$

$$x_D^{(3)} = \left(\frac{1.0}{1.7}\right)^{2.3691}\frac{0.8350}{0.6}0.36 = 0.1425$$

The Underwood equation (Equation 4.11) is solved for the value of ϕ, where $1.7 \le \phi \le 1.0$. The value of ϕ is found to be 1.1214. By substituting this value in the following equation, we can get the value of $R_{min\ u}$.

$$R_{min\ u} = \frac{1.7 \times 0.8350}{1.7 - 1.1214} + \frac{1.16 \times 0.0225}{1.16 - 1.1214} + \frac{1.0 \times 0.1425}{1.0 - 1.1214} - 1$$

$$= 0.9557$$

Given that $N = 10 + 1$ (for the reboiler) and $C_1 = 2.3691$, the value of Y in Gilliland's correlation is found to be

$$Y = \frac{11 - 2.3691}{11 + 1} = 0.7192$$

By substituting the value of Y in Gilliland's correlation, Equation 4.13, the value of X is found to be 0.00957. Since $R = 1$,

$$R_{min\ g} = R - X(R + 1) = 0.9809$$

Therefore,

$$G_c = \frac{R_{min\ u}}{R} - \frac{R_{min\ g}}{R} = -0.0252 \ge 0.0001$$

Since G_c is not within the tolerance, we need to iterate on C_1. Iteratively the value of C_1 which will satisfy the tolerance is found to be 2.4147 and hence the values of $x_D^{(i)}, i = 1, 2, 3$ are 0.8382, 0.0222, and 0.1396.

b) Figure 4.4 shows the comparison of the shortcut method results to those obtained from the rigorous model for $LK = 1$ and $HK = 3$. It can be easily seen that the results match very closely.

Example 4.2: A quaternary equimolar feed of 100 moles containing components with relative volatilities of 2.0, 1.5, 1.0, and 0.5. is to be distilled. The distillation column available for the operation has ten theoretical plates.

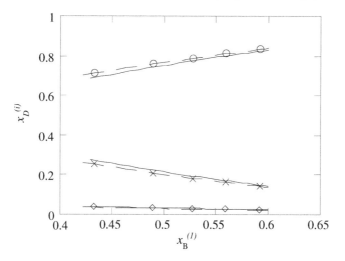

FIGURE 4.4
Comparison of the Shortcut Method with the Rigorous Model for Example 4.1b

a) Simulate the first time step operation if the column is operating in a constant reflux mode with a reflux ratio equal to 1.0. Assume $k = 1$, $LK = 1$, and $HK = 3$. Assume the initial value of C_1 equal to 2.3691.

b) Change the value of LK and HK where $LK = 1$ and $HK = 4$ and compare the results with a).

c) Simulate a 1 hr operation of the column if the vapor boilup rate V is 100 mole/hr and compare with the results of the simulation using the rigorous model presented in Chapter 3. Assume $LK = 1$ and $HK = 3$.

d) Change the value of $LK = 1$ and $HK = 4$ and compare the results with c).

Solution:

a) Assuming that initially $C_1 = 2.3691$, from the Hengestebeck–Geddes equation the value of $x_D^{(1)}$ is:

$$
\begin{aligned}
x_D^{(1)} &= 1/\left(\frac{2.0}{2.0}\right)^{2.3691}\frac{0.25}{0.25} + \left(\frac{1.5}{2.0}\right)^{2.3691}\frac{0.25}{0.25} \\
&+ \left(\frac{1.0}{2.0}\right)^{2.3691}\frac{0.25}{0.25} + \left(\frac{0.5}{2.0}\right)^{2.3691}\frac{0.25}{0.25} \\
&= 0.5758
\end{aligned}
$$

Therefore, $x_D^{(i)}$, $i = 2, 3, 4$ are given by:

$$
x_D^{(2)} = \left(\frac{1.5}{2.5}\right)^{2.3691}\frac{0.5758}{0.25}0.25 = 0.2912
$$

$$x_D^{(3)} = \left(\frac{1.0}{2.5}\right)^{2.3691} \frac{0.5758}{0.25} 0.25 = 0.1114$$

$$x_D^{(4)} = \left(\frac{0.5}{2.5}\right)^{2.3691} \frac{0.5758}{0.25} 0.25 = 0.0216$$

The Underwood equation (Equation 4.11) is solved for the value of ϕ, where $2.0 \le \phi \le 1.0$. The value of ϕ is found to be 1.1639. By substituting this value in the following equation, we can get the value of $R_{min\ u}$.

$$
\begin{aligned}
R_{min\ u} &= \frac{2.0 \times 0.5758}{2.0 - 1.1639} + \frac{1.5 \times 0.2912}{1.5 - 1.1639} + \frac{1.0 \times 0.1114}{1.0 - 1.1639} \\
&+ \frac{0.5 \times 0.0216}{0.5 - 1.1639} - 1 \\
&= 0.9806
\end{aligned}
$$

Given that $N = 10 + 1$ (for the condenser) and $C_1 = 2.3691$, the value of Y in Gilliland's correlation is found to be

$$Y = \frac{11 - 2.3691}{11 + 1} = 0.7192$$

By substituting the value of Y in Gilliland's correlation, Equation 4.13, the value of X is found to be 0.0094. Since $R = 1$,

$$R_{min\ g} = R - X(R + 1) = 0.9815$$

Therefore,

$$G_c = \frac{R_{min\ u}}{R} - \frac{R_{min\ g}}{R} = -0.0009 \le 0.0001$$

Since the value of G_c is found to be within the tolerance (0.0001), the loop has converged and the values of $x_D^{(i)}, i = 1, 2, \ldots, 4$ for this time step are found to be 0.5758, 0.2912, 0.1114, and 0.0216.

b) If we change the value of light key and heavy key components, it will only change the calculation for $R_{min\ u}$. For $LK = 1$ and $HK = 4$, the value of ϕ is found to be 0.5907, which results in

$$
\begin{aligned}
R_{min\ u} &= \frac{2.0 \times 0.5758}{2.0 - 0.5907} + \frac{1.5 \times 0.2912}{1.5 - 0.5907} + \frac{1.0 \times 0.1114}{1.0 - 0.5907} \\
&+ \frac{0.5 \times 0.0216}{0.5 - 0.5907} - 1 \\
&= 0.4509
\end{aligned}
$$

and hence

$$G_c = \frac{R_{min\ u}}{R} - \frac{R_{min\ g}}{R} = 0.5300 > 0.0001$$

Since G_c is not within the tolerance, we need to iterate on C_1. Iteratively the value of C_1 which will satisfy the tolerance is found to be 0.00125 and hence the values of $x_D^{(i)}, i = 1, 2, \ldots, 4$ are 0.8439, 0.1440, 0.0119, and 0.0002. Note that the values are different from those obtained in a).

c) Figure 4.5 shows the comparison of the shortcut method results to those obtained from the rigorous model for $LK = 1$ and $HK = 3$. It can be easily seen that the results match very closely.

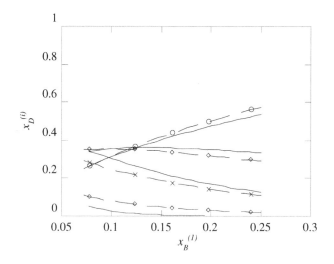

FIGURE 4.5
Comparison of the Shortcut Method with the Rigorous Model for Example 4.2c

(d) Figure 4.6 shows the comparison of the shortcut method results with $LK = 1$ and $HK = 3$ to those with $LK = 1$ and $HK = 4$. Although the same shortcut method is used for the same column, because the light key and heavy key components are inappropriately defined, the converged results for the two cases are different. Therefore, the user should be careful in choosing the appropriate values for the light key and heavy key for successful use of this method.

Variable Reflux:

1. Assume C_1

2. Since $x_D^{(k)}$ is constant, the following equation, resulting from summation of distillate composition of all the components is used to iterate on the value of C_1.

$$G_v = \sum_{i=1}^{n} \left(\frac{\alpha_i}{\alpha_k} \right)^{C_1} x_B^{(i)} - \frac{x_B^{(k)}}{x_D^{(k)}} \qquad (4.17)$$

Find the value of G_v using the above equation.

3. If G_v is approximately zero then go to the next step. Otherwise, use the Newton-Raphson Method to obtain the new value of C_1 and repeat step 2.

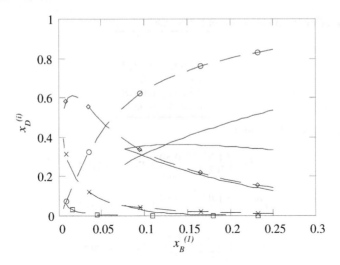

FIGURE 4.6
Comparison of the Shortcut Method with Different Key Components for Example 4.2d

4. Substitute this converged value of C_1 in the Hengestebeck–Geddes equation and find the distillate composition, $x_D^{(i)}$, $i = 1, 2, \ldots, n$.

5. Solve the Underwood equations by substituting the values of the distillate and the bottom composition for R_{min}.

6. Equate C_1 to N_{min} and calculate the value of R from Gilliland's correlation using the calculated values of N_{min} and R_{min} for specified N.

Example 4.3: Use the same column as in Example 4.2. Simulate the first time step operation if the column is operating in a variable reflux mode with the key component distillate composition $x_D^{(1)}$ equal to 0.80, which should remain constant throughout the operation. Assume $LK = 1$ and $HK = 3$.

Solution: Assume that initially $C_1 = 2.3691$, and given $x_D^{(1)} = 0.80$, from Equation 4.17 the value of G_v is found to be

$$
\begin{aligned}
G_v &= \frac{0.25}{0.80} - \left(\frac{2.0}{2.0}\right)^{2.3691} 0.25 + \left(\frac{1.5}{2.0}\right)^{2.3691} 0.25 + \left(\frac{1.0}{2.0}\right)^{2.3691} 0.25 \\
&\quad + \left(\frac{0.5}{2.0}\right)^{2.3691} 0.25 \\
&= 0.1217
\end{aligned}
$$

Since G_v is not within the tolerance, we have to iterate on the value of C_1. The following iterative scheme based on the Newton-Raphson method is used to obtain the new value of C_1.

$$
C_{1\ new} = C_{1\ old} - \frac{G_v}{dG_v/dC_1}
$$

The iteration summary is provided below.

C_1	G_v
2.3691	0.1217
3.8372	0.0391
4.8755	0.0078
5.2004	0.0005

Hence, $x_D^{(i)}$, $i = 2, 3, 4$ are:

$$x_D^{(2)} = \left(\frac{1.5}{2.5}\right)^{5.2}\frac{0.80}{0.25}0.25 = 0.1792$$

$$x_D^{(3)} = \left(\frac{1.0}{2.5}\right)^{5.2}\frac{0.80}{0.25}0.25 = 0.0217$$

$$x_D^{(4)} = \left(\frac{1.0}{2.5}\right)^{5.2}\frac{0.80}{0.25}0.25 = 0.0006$$

The Underwood equation (Equation 4.11) is solved for $2.0 \leq \phi \leq 1.0$. The value of ϕ is found to be 1.1639. By substituting this value in the following equation, we can get the value of $R_{min\ u}$.

$$\begin{aligned}R_{min\ u} &= \frac{2.0 \times 0.8000}{2.0 - 1.1639} + \frac{1.5 \times 0.1780}{1.5 - 1.1639} + \frac{1.0 \times 0.0214}{1.0 - 1.1639}\\ &+ \frac{0.5 \times 0.0006}{0.5 - 1.1639} - 1\\ &= 1.577\end{aligned}$$

Given that $N = 10 + 1$ (for the condenser) and $C_1 = 5.22$, the value of Y in Gilliland's correlation is found to be

$$Y = \frac{11 - 5.22}{11 + 1} = 0.4813$$

By substituting the value of Y in Gilliland's correlation, Equation 4.13, the value of X is found to be 0.1761. Substituting $R_{min} = 1.5770$ and $X = 0.1761$,

$$R = \frac{X - R_{min}}{1 - X} = 2.1277$$

The above examples illustrated the single time step procedure and the complete simulation procedure involved in the shortcut method. This method has been tested and compared with rigorous simulation models (described in Chapter 3) using several examples. It has been found to be extremely efficient and reasonably accurate for thermodynamically nearly ideal mixtures and for columns with negligible holdup effects. It has also been shown that with the shortcut method it is possible to extract global properties of the batch column in terms of feasible region or inequality constraints, especially for design variables like the number of plates N and the reflux ratio R. These are very useful properties of the shortcut method. The identification of bounds on design parameters is very handy in design, optimization, and control problems and Chapter 5 will discuss the feasible operating regions in detail. This method has also been extended to complex mixtures containing binary and ternary azeotropes (Diwekar, 1991; Kalagnanam and Diwekar, 1993). Chapter 7 focuses on these complex systems and describes these extensions.

4.3   Modified Shortcut Method

The shortcut method presented earlier demonstrated for the first time the power of aggregation in the context of batch distillation. The method represents the first level in the hierarchy of models for batch distillation. Next in the hierarchy is the modified shortcut method, which takes into consideration holdup effects by including additional dynamics in the shortcut model.

The assumption of negligible holdup is reasonable for certain applications, and the transient profiles generated by the shortcut method (dashed lines marked with symbols) approximate the column dynamics accurately, as shown in Figures 4.7 and 4.8. Holdup effects do play an important role in several other batch distillation columns (Figures 4.9 and 4.10).

Holdup can affect the performance of a batch distillation column in two basic ways, namely, the dynamic flywheel effect, and the steady state capacitance effect. The flywheel effect can be characterized by the parameter $\tau = holdup/(reflux \times distillate\ rate)$. For large τ, the initial composition profile predicted by the zero holdup models such as the shortcut method departs significantly from the results of a rigorous model. The capacitance effect is observed at the end of the total reflux condition, when the given charge distributes itself throughout the column. Associated with the capacitance effect is the concept of equilibration time. The shortcut method assumes instant equilibration at the total reflux condition, and the estimation of equilibration time is not possible using the shortcut method. However, for ideal binary system quick estimates of equilibration time can be obtained from shortcut correlations (Ellerbe, 1979; Sørensen, 1994).

The modified shortcut method (Diwekar, 1994) is based on the compartmental modeling approach proposed for continuous distillation by Benallou et al. (1986). The compartmental modeling approach is based on the assumption that the number of plates in a distillation column can be lumped to form an equivalent plate whose dynamic response resembles that of the dynamic response of the compartment. Since holdup represents the accumulation terms that govern the dynamics of the system, the compartment is assigned a mass holdup equivalent to the sum of individual holdups and the composition x_s of the sensitive plate. Figure 4.11 shows the batch distillation column, which consists of several compartments. The following equations describe the overall dynamics of each compartment in terms of the compartmental input and

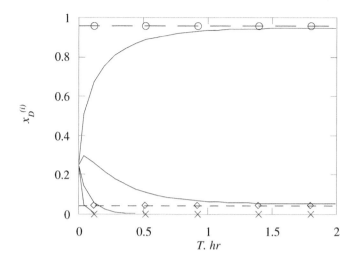

FIGURE 4.7
Transient Profiles for Batch Distillation, Total Reflux Operation for a Quaternary System, 22% holdup, $N = 10$, $\alpha = 2.0, 1.5, 1.0, 0.5$

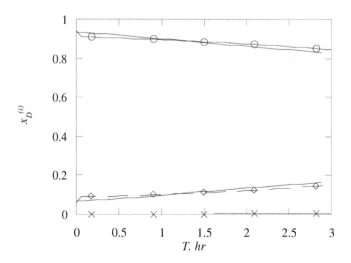

FIGURE 4.8
Transient Profiles for Batch Distillation, Finite Reflux Operation for a Quaternary System, 25% holdup, $N = 10$, $\alpha = 2.0, 1.5, 1.0, 0.5$; $R = 8.0$

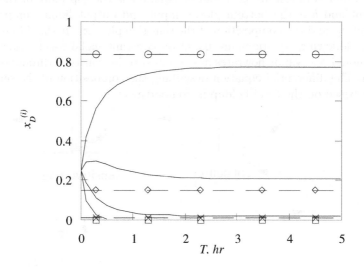

FIGURE 4.9
Capacitance Effect of Holdup for a Quaternary System, 25% holdup, $N = 5$, $\alpha = 2.0, 1.5, 1.0, 0.5$

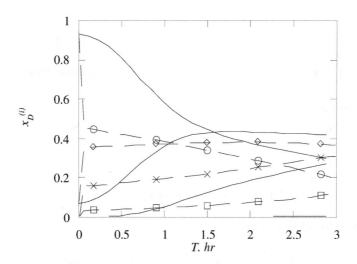

FIGURE 4.10
Flywheel Effect of Holdup for a Quaternary System, 25% holdup, $N = 10$, $\alpha = 2.0, 1.5, 1.0, 0.5$; $R = 1.0$

output vapor composition y_{b+1} and y_t (where t is the top plate of the compartment and b is the bottom plate), input and output liquid composition x_{t-1} and x_b, and the composition of the sensitive plate x_s. It should be noted that the following equations are based on constant molal assumptions and, hence, liquid and vapor flowrates are assumed to be constant throughout the column. The differential equation describing the composition of the sensitive plate is based on the total holdup in compartment H_c.

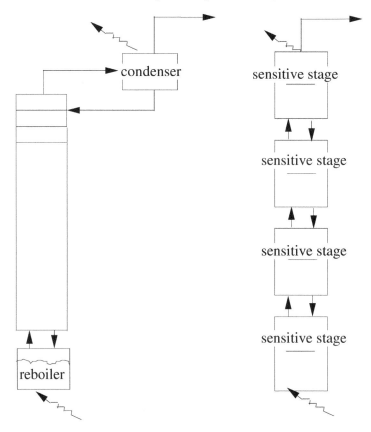

FIGURE 4.11
Compartmental Model for Batch Distillation

Compartment dynamics:

$$H_c \frac{dx_s}{dt} = V(y_{b+1} - y_t) + L(x_{t-1} - x_b) \qquad (4.18)$$

where

$$H_c = \sum_{i=t}^{b} H_i$$

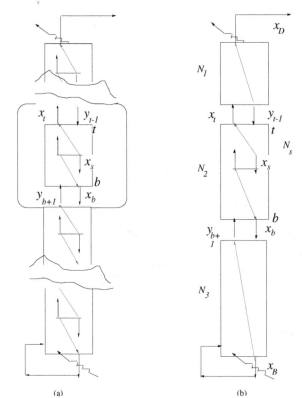

FIGURE 4.12
Calculation Envelopes for the Compartmental Model

The values of the compositions leaving each compartment (y_t and x_b) can be found algebraically using the steady state relationship either with the bottom composition x_B or with the top composition x_D. This can be visualized as material balance envelopes, shown in Figure 4.12. The steady state compartmental balance is obtained with the shortcut method.

This approach has been shown to capture the capacitance effect of holdup very easily. However, it was observed that for the flywheel effect, the use of the shortcut method for compartmental balances results in multiple solutions (Diwekar, 1992b), as shown in Figure 4.13. This problem is attributed to the numerical instabilities when the reflux ratio approaches the minimum reflux condition. Although attempts to circumvent this problem were proposed, this approach is not very robust for the flywheel effect. Hence, we will only present this approach in the context of the capacitance effect.

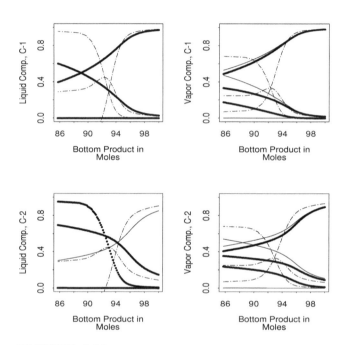

FIGURE 4.13

Multiple Solutions in the Modified Shortcut Method for the Flywheel Effect (solid lines, rigorous model results; C-1, compartment 1 and C-2, compartment 2)

Capacitance Effect and Equilibration Time

To demonstrate the compartmental model concept the column shown in Figure 4.12a is decomposed into three compartments. Figure 4.12b shows these three compartments where the number of plates associated with each compartment (top, middle, and bottom) is denoted by N_1, N_2, and N_3, respectively. The compartmental balances are represented by the calculation envelopes around each compartment. It can be seen from the figure that the liquid composition of the stream reaching the top plate of the compartment x_{t-1} is calculated using the distillate composition, the vapor composition of the stream reaching the bottom plate of the compartment y_b is a function of the still composition, and the sensitive plate composition provides the estimates for the composition of the vapor leaving the compartment top and the liquid leaving the bottom of the compartment (through the vapor composition leaving the sensitive plate). The shortcut method equations are used to solve these material balance envelopes.

At the total reflux condition, the compartmental shortcut method reduces to using the Hengestebeck–Geddes equation and the Fenske equation, where the design variable number of plates (N) is the same as the minimum number of plates (N_{min}). Likewise, the Hengestebeck–Geddes constant C_1 is equal to the number of plates. The following system of equations describes the dynamics of batch distillation operating at the total reflux condition using the three compartment model.

Dynamics of the last compartment (with the reboiler as the sensitive plate):

$$\frac{dx_B^{(i)}}{dt} = \frac{V(x_b^{(i)} - y_{b+1}^{(i)})}{B + \sum_{i=N_2+1}^{N_3} H_i} \qquad x_{B0}^{(i)} = x_F^{(i)} \tag{4.19}$$

Dynamics of the middle compartment(s):

$$\frac{dx_s^{(i)}}{dt} = \frac{V(y_{b+1}^{(i)} - y_t^{(i)}) - V(x_b^{(i)} - x_{t-1}^{(i)})}{\sum_{i=N_1+1}^{N_2} H_i} \qquad x_{s0}^{(i)} = x_F^{(i)} \tag{4.20}$$

Dynamics of the first compartment (with the condenser as the sensitive plate):

$$\frac{dx_D^{(i)}}{dt} = \frac{V y_t^{(i)} - V x_{t-1}^{(i)}}{H_D + \sum_{i=1}^{N_1} H_i} \qquad x_{D0}^{(i)} = x_F^{(i)} \tag{4.21}$$

where s is the subscript used for the sensitive plate in the middle compartment (counted from the top of the compartment). The steady state material balance envelopes can be obtained using the following shortcut method equations.

Composition of the liquid stream leaving the top compartment:

$$x_{t-1}^{(i)} = \frac{(\frac{\alpha_i}{\alpha_1})^{-N_1} x_D^{(i)}}{\sum_{k=1}^{n}(\frac{\alpha_k}{\alpha_1})^{-N_1} x_D^{(k)}} \tag{4.22}$$

Composition of the vapor stream leaving the middle compartment(s):

$$y_t^{(i)} = \frac{\left(\frac{\alpha_i}{\alpha_1}\right)^{N_s} x_s^{(i)}}{\sum_{k=1}^{n}\left(\frac{\alpha_k}{\alpha_1}\right)^{N_s} x_s^{(k)}} \tag{4.23}$$

Composition of the liquid stream leaving the middle compartment(s):

$$x_b^{(i)} = \frac{\left(\frac{\alpha_i}{\alpha_1}\right)^{-(N_2-N_s)} x_s^{(i)}}{\sum_{k=1}^{n}\left(\frac{\alpha_k}{\alpha_1}\right)^{-(N_2-N_s)} x_s^{(k)}} \tag{4.24}$$

Composition of the vapor stream leaving the bottom compartment:

$$y_{b+1}^{(i)} = \frac{\left(\frac{\alpha_i}{\alpha_1}\right)^{N_3+1} x_B^{(i)}}{\sum_{k=1}^{n}\left(\frac{\alpha_k}{\alpha_1}\right)^{N_3+1} x_B^{(k)}} \tag{4.25}$$

Using the dynamics of the compartments given by Equations 4.19-4.21, along with the quasi-steady state equations (Equations 4.22-4.25), the transient behavior of the total reflux condition can be simulated. Note that the selection of the number of plates for each compartment and the sensitive plate in each compartment is very important to obtain an accurate response of the system.

Example 4.4: A four-component mixture of an equimolar feed of 100 mole with relative volatilities of 2.0, 1.5, 1.0, and 0.5 is to be distilled in a five-plate column. Assume that the column has 25% holdup. Simulate the total reflux transient operation of the column in the total reflux mode.

a) The compartmental model consists of three compartments of $N_1 = 4$, $N_2 = 1$, and $N_3 = 0$ with the sensitive plate $N_s = 1$. Compare the results with the rigorous simulation results obtained in Chapter 3.

b) What is the equilibration time?

Solution:

a) Figure 4.14 shows the comparison of the distillate composition profiles obtained by the compartmental model and the rigorous model. For these models with $N_1 = 4$, $N_2 = 1$, and $N_3 = 0$, and the sensitive plate $N_s = 1$, the results compare very well.

b) The equilibration time is found to be approximately 4.7, which is comparable to the equilibration time obtained by the rigorous model simulation ($T_{eq} = 4.7$).

4.4   Collocation-Based Model

The shortcut method is applicable to nearly ideal (or ideal azeotropic) systems with a constant molal overflow assumption and columns with negligible holdup

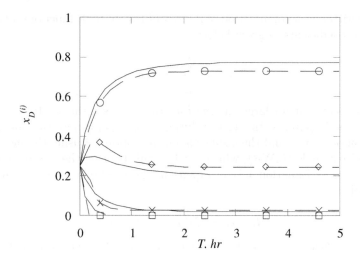

FIGURE 4.14
Comparison of the Compartmental Model in Example 4.4a with the Rigorous
Model

effects. It has a minimum number of differential equations and a few algebraic
equations to describe the complete dynamics. The modified shortcut method
considers holdup effects and has additional dynamic and algebraic equations
but is also based on the assumption of constant molal overflow and nearly ideal
thermodynamics. The next simplified model in the hierarchy is the reduced
order model based on the collocation approach.

The collocation approach was first proposed in the context of continuous
staged separation processes by Chao and Joseph (1983a, 1983b) followed by
a number of articles on this topic by Stewart et al. (1985), Srivastava and
Joseph (1984, 1985, 1987). Diwekar (1988) extended this approach to obtain
the reduced order model for batch distillation discussed in this section.

The collocation approach to model reduction is based on approximating
the column stage variables using polynomials rather than a discrete function
of stages. The polynomial approximation is obtained using the theory of orthogonal collocation. To understand this concept, we will first describe why
orthogonal collocation is useful, what orthogonal collocation is, and its significance in solving a set of differential equations.

Why Orthogonal Collocation?

Consider a general nonlinear differential equation:

$$D(y) \;=\; f(x,y) \tag{4.26}$$

where $D(y)$ is the differential operator. In general, the solution of this equation can be written as a combination of a series of known basis functions $\theta_i(x)$ and unknown coefficients a_i given below.

$$y_s = \sum_{i=1}^{nn} a_i \theta_i(x) \tag{4.27}$$

We expect that for large values of nn ($nn \to \infty$), the solution in Equation 4.27 will approach the exact solution. The $\theta_i(x)$ denote the polynomial basis functions of x and the coefficients a_i are obtained using the method of weighted residuals (MWR), which involves substituting the approximate solution given by Equation 4.27 in Equation 4.26 to obtain a residual function given below.

$$R(x, a) = D(y_s) - f(x, y_s) \tag{4.28}$$

The coefficients a_i are determined by minimizing the residual function $R(x, a)$ over the desired range of the independent variable x (for example, over $0 \leq x \leq 1$) along with a choice of the weighting function, $w(x)$. The residual minimization function may be written as follows:

$$\int_0^1 R(x, a)w(x)dx = 0 \tag{4.29}$$

Equation 4.29 describes the generalized form of all the methods of weighted residuals. The specific choice of the weighting function depends on the method used. The following list provides a few of these methods.

- Method of Least Squares

- Method of Galerkin (Galerkin, 1915)

- Method of Collocation (Frazer et al., 1937; Lanczos, 1938)

The least squares method was originally proposed by Gauss in 1795 in the context of parameter estimation. The weighting function in the least squares method is $dR(x, a)/da$ which results in:

$$\int_0^1 R(x, a)\frac{dR(x, a)}{da}dx = 0 \tag{4.30}$$

$$\text{Min} \int_0^1 (R(x, a))^2 dx \tag{4.31}$$

The least squares method can become very cumbersome for the solution of differential equations.

The Galerkin method, on the other hand, uses dy_s/da as the weighting function. The Galerkin method chooses an orthogonal class of functions and thus forces the residual to be zero by making it orthogonal to each member of a complete set (in the limit as $nn \to 0$). The Galerkin method is one of the best known approximate methods of weighted residuals.

The collocation method uses a delta function as a weighting function, given by

$$w(x) = \delta(x - x_k), k = 1, 2, \ldots, ncol \tag{4.32}$$

where the x_k are $ncol$ points between 0 and 1. This is equivalent to saying that the residual goes to zero at every collocation point.

$$R(a, x_k) = 0, k = 1, 2, \ldots, ncol \tag{4.33}$$

Now if the locations of the collocation points are the zeros of an orthogonal polynomial, the solution of Equation 4.27 approaches the Galerkin approximation (Villadsen, 1970), which has been shown to be the most accurate method of weighted residuals. The difference is that the Galerkin method is not suitable for machine computation while orthogonal collocation is easily programmable and is as accurate as the Galerkin method.

In the orthogonal collocation method, the collocation points are taken as roots of orthogonal polynomials. This procedure was first advanced by Lanczos (1938) and was developed further for the solution of ordinary differential equations using the Chebyshev series. These applications were primarily for initial-value problems. In 1967, Villadsen and Stewart made a major advance when they developed orthogonal collocation for boundary-value problems. They chose the functions to be sets of orthogonal polynomials that satisfied the boundary conditions while the roots of the polynomials gave the collocation points. Thus, the choice of the collocation points is no longer arbitrary and the lower-order collocation results are more dependable. A major simplification is that the solution can be derived in terms of the coefficients in the function and the values of the solution at the collocation points. The whole problem is then reduced to a set of matrix equations, which are easily generated and solved numerically (Finlayson, 1972).

In orthogonal collocation, the trial function given in Equation 4.27 is written in terms of linear combinations of orthogonal polynomials $P_m(x)$ of the order 1 to $m + 1$, with P_0 as the starting point.

$$y_s = \sum_{i=1}^{m} c_i P_{m-1}(x) \tag{4.34}$$

There are a number of different kinds of orthogonal polynomials one can use, including continuous polynomials like Lagrange or Legendre polynomials (Villadsen and Michalesen, 1978; Finlayson, 1972; Chao and Joseph, 1983a, b), and discrete ones, such as Hahn's polynomial (Stewart et al., 1985). The orthogonality property allows one to obtain the roots of the polynomial x_i, $i = 1, 2, \ldots, m - 1$. Since orthogonal polynomials are also formed by linear combination of x or x^2 (for simplicity we can take the example of polynomials in x), Equation 4.34 can be rewritten at each collocation point in terms of new coefficients d_i

$$y_j = \sum_{i=1}^{m} d_i x_{i-1}; \qquad j = 1, 2, \ldots, m-1 \tag{4.35}$$

Differentiating the above equation results in:

$$\frac{dy}{dx}\Big|_{x_j} = \sum_{i=1}^{m} d_i \frac{dx_{i-1}}{dx}\Big|_{x_j}; \qquad j = 1, 2, \ldots, m-1 \tag{4.36}$$

$$\frac{d^2 y}{dx^2}\Big|_{x_j} = \sum_{i=1}^{m} d_i \frac{d^2 x_{i-1}}{dx^2}\Big|_{x_j}; \qquad j = 1, 2, \ldots, m-1 \tag{4.37}$$

Substituting the value of d_i from Equation 4.35 in the above equations:

$$\frac{dy}{dx}\Big|_{x_j} = \sum_{i=1}^{m} A_{ji} y_i; \qquad j = 1, 2, \ldots, m-1 \tag{4.38}$$

$$\frac{d^2 y}{dx^2}\Big|_{x_j} = \sum_{i=1}^{m} B_{ji} y_i; \qquad j = 1, 2, \ldots, m-1 \tag{4.39}$$

To evaluate the integral, a simple quadrature formula can be used.

$$\int_{o}^{1} y\, dx = \sum_{i=1}^{m} W_i f(x_i, y_i) \tag{4.40}$$

where

$$[A_{ji}] = [\tfrac{dx_{i-1}}{dx}\big|_{x_j}][x_{i-1}]^{-1}$$

$$[B_{ji}] = [\tfrac{d^2 x_{i-1}}{dx^2}\big|_{x_j}][x_{i-1}]^{-1}$$

$$[W_i] = [f(x_j, y_j)][x_{i-1}]^{-1}$$

From Equations 4.38 and 4.39, one can easily see that the differential operator can be replaced by a linear combination of the solution y at the collocation points $j = 1, 2, \ldots, m-1$ and, hence, the ordinary differential equations are reduced to a set of algebraic equations as given below.

$$\sum_{i=1}^{m} D_{ji} y_i = f(x_j, y_j); \qquad j = 1, 2, \ldots, m-1 \tag{4.41}$$

With this technique, the partial differential equations can be reduced to a set of ordinary differential equations (or a set of algebraic equations). The following example will illustrate the orthogonal collocation technique for the initial-value problem.

Example 4.5: Solve the following differential equations using the orthogonal collocation technique for different numbers of collocation points and compare it with the analytical solution in the range 0 to 1.

$$\frac{dy}{dx} = 10x, \qquad y(0) = 0.0$$

For this purpose, use the shifted Legendre polynomials whose roots and matrices are provided below.

Roots and Matrices of Shifted Legendre Polynomials (Finlayson, 1972)

$$N = 1 \quad x = \begin{bmatrix} 0 \\ 0.5000 \\ 1 \end{bmatrix} \quad W = \begin{bmatrix} \frac{1}{6} \\ \frac{2}{3} \\ \frac{1}{6} \end{bmatrix}$$

$$A = \begin{bmatrix} -3 & 4 & -1 \\ -1 & 0 & 1 \\ 1 & -4 & 3 \end{bmatrix} \quad B = \begin{bmatrix} 4 & -8 & 4 \\ 4 & -8 & 4 \\ 4 & -8 & 4 \end{bmatrix}$$

Solution: The analytical solution of the above differential equation is given by

$$\int dy = \int 10x \, dx = 5x^2 + c$$

Since $y(0) = 0$ the above equation can be simplified as:

$$y = 5x^2$$

In the collocation method the solution is obtained at collocation points ($x_1 = 0$; $x_2 = 0.5$; $x_3 = 1.0$). From the above analytical expression the values of y at the three collocation points are given by

$$y_1 = 0; \quad y_2 = 1.25; \quad y_3 = 5.0$$

Now we will solve the problem using the collocation method. From Equation 4.38 the following equation can be derived

$$\sum_{i=1}^{ncol+2} A_{ij}y_j = 10x_j; \qquad j = 2, \ldots, ncol + 2$$

Substituting the values of x_j in the above equations results in the following equations in y_j.

$$\begin{aligned} -3y_1 + 4y_2 - 1y_3 &= 10x_1 \\ -1y_1 + 0y_2 + 1y_3 &= 10x_2 \\ 1y_1 - 4y_2 + 3y_3 &= 10x_3 \end{aligned}$$

The simultaneous solution of the above equations provides

$$y_1 = 0; \quad y_2 = 1.25; \quad y_3 = 5.0$$

which is exactly same as the analytical solution.

Order Reduction in Batch Distillation

In the last section, we saw that the orthogonal collocation technique can change partial differential equations to ordinary differential or algebraic equations, and ordinary differential equations (ODEs) to a set of algebraic equations. In the case of batch distillation columns, we encounter ordinary differential equations, and orthogonal collocation can be used to reduce this system of ODEs into nonlinear equations. However, it is not always advantageous to convert ODEs to nonlinear algebraic equations. In fact, often it is preferable to use ODEs in the place of nonlinear equations for model reduction (vanDongen and Doherty, 1985), in light of which the use of orthogonal collocation may seem unnecessary. However, we will show that instead of using the orthogonal collocation technique to reduce the ODEs to nonlinear algebraic equations, one can use it to reduce the order of ODEs, keeping the dynamic nature (ODEs) of the model as it is.

In the context of continuous stage separation processes, Wong and Luus (1980) proposed the use of orthogonal collocation to reduce the model order of the system of ordinary differential equations. The concept is derived from the use of partial differential equations to approximate the column dynamics, whereas the stagewise column can be approximated as a distributed system in which the composition and flow profiles can be represented by continuous variables along the length of the column. Chao and Joseph (1983a,b), and Srivastava and Joseph (1985, 1987) discussed specific aspects of the collocation approach such as choosing the number of collocation points, the problem of steep profiles, etc. They showed that the method can be applied directly without the need to generate partial differential equations. They used Jacobi polynomials, which are continuous orthogonal polynomials. Stewart et al. (1985) used discrete orthogonal polynomials (Hahn polynomials, which are discrete analogs of Jacobi polynomials) and showed that these characterize the dynamics of a staged column more accurately, as they capture the discrete nature of the process.

The applicability of such an order reduction for batch distillation was explored by Diwekar and co-workers (Diwekar, 1988). The aim was to test the utility of this approach. It was found that this order reduction procedure can be successfully applied to batch distillation, and that the reduction is significant. Furthermore, the problems of steep profiles, namely, that of lighter than light key (LLK) and heavier than heavy key (HHK) that made this order reduction approach less attractive in the case of continuous distillation (Srivastava and Joseph, 1987) are not observed in the case of batch distillation. The application of the collocation approach for batch distillation problems is described below. For ease of understanding, the discussion starts with a partial differential approximation of the column.

Figure 4.15 shows the complete dynamic model of the batch distillation column described in Chapter 3, with the difference being that the tray indexing of the liquid and vapor variables is changed. In the case of the conventional model of the batch distillation column presented in Chapter 3, the liquid and vapor leaving tray j are denoted by L_j and V_j, respectively. In the case of the collocation-based model, the vapor entering the plate and the liquid leaving the plate are represented by the subscript j. Figure 4.15 shows these differences. This type of notation is convenient for converting ordinary differential equations to partial differential equations in time and space, and also for the order reduction procedure. The complete dynamic equations of the column given in Table 3.1 are rewritten below, using the new nomenclature shown in Figure 4.15. These equations are based on the same assumptions as described in Chapter 3 for the equations in Table 3.1.

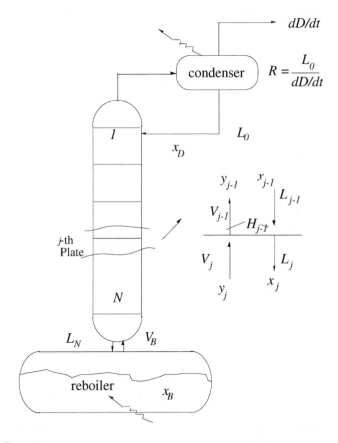

FIGURE 4.15
Schematic of the Batch Distillation Column for the Reduced Order Collocation Model

Condenser and accumulator dynamics:

$$\frac{dx_D^{(i)}}{dt} = \frac{V_0}{H_D}(y_0^{(i)} - x_D^{(i)}), i = 1, 2, \ldots, n \qquad (4.42)$$

Plate dynamics:

$$\frac{dx_j^{(i)}}{dt} = \frac{1}{H_j}[V_j y_j^{(i)} + L_{j-1} x_{j-1}^{(i)} - V_{j-1} y_{j-1}^{(i)} - L_j x_j^{(i)}]$$

$$i = 1, 2, \ldots, n; j = 1, 2, \ldots, N \qquad (4.43)$$

Reboiler dynamics:

$$\frac{dx_B^{(i)}}{dt} = \frac{1}{B}[L_N(x_N^{(i)} - x_B^{(i)}) - V_N(y_N^{(i)} - x_B^{(i)})] \qquad (4.44)$$

Heat balance and flowrate calculations at the top of the column:

$$L_0 = R\frac{dD}{dt}$$

$$V_0 = \frac{dD}{dt}(R+1) + \delta_t H_D \qquad (4.45)$$

On the plates:

$$L_j = V_j + L_{j-1} - V_{j-1} - \delta I_j; \quad j = 1, 2, \ldots, N$$

$$V_{j-1} = \frac{1}{J_j - I_j}[V_{j-1}(J_{j-1} - I_j) + L_{j-1}(I_{j-1} - I_j)$$

$$+ \quad H_j \delta I_j]; \quad j = 1, 2, \ldots, N+1 \qquad (4.46)$$

At the bottom of the column

$$\frac{dB}{dt} = L_N - V_N \qquad (4.47)$$

Heat duty calculations:
Condenser duty

$$Q_D = V_0(J_0 - I_0) - H_D \delta_t I_0 \qquad (4.48)$$

Reboiler duty

$$Q_B = V_N(J_N - I_N) - L_N(I_N - I_B) + B\delta_t I_B \qquad (4.49)$$

The vapor liquid equilibria equations for each plate and reboiler and condenser in the general form are

$$y_{j-1}^{(i)} = \frac{\alpha_i x_j^{(i)}}{(\sum_{k=1}^{n} \alpha_k x_j^{(k)})}, \quad i = 1, 2, \ldots, n; j = 1, 2, \ldots, N \quad (4.50)$$

$$y_N^{(i)} = \frac{\alpha_i x_B^{(i)}}{(\sum_{k=1}^{n} \alpha_k x_B^{(k)})}, \quad i = 1, 2, \ldots, n \qquad (4.51)$$

where $\alpha_i = f(x_j^{(l)}, l = 1, 2, \ldots, n, \ TE_j, P_j)$

To understand the concept of applying orthogonal collocation, we will initially neglect the heat balance and assume constant molal overflow. This means using the following equations for flowrate calculations.

$$
\begin{aligned}
L_0 &= L_j = L, & j &= 1, 2, \ldots, N \\
V_0 &= V_j = V, & j &= 1, 2, \ldots, N \\
H_j &= H, & j &= 1, 2, \ldots, N
\end{aligned}
\tag{4.52}
$$

In order to generate the partial differential equations in continuous space $z = 0$ to $z = 1$, we will assume that x and y are smooth functions of the distance variable z, leading to the following approximation.

$$
x_{j-1}^{(i)} = x_j^{(i)} - \frac{\partial x_j^{(i)}}{\partial z} \triangle z
\tag{4.53}
$$

$$
y_{j-1}^{(i)} = y_j^{(i)} - \frac{\partial y_j^{(i)}}{\partial z} \triangle z
\tag{4.54}
$$

where $\triangle z$ is the spacing between the plates given by $\triangle z = \frac{1}{N}$. Substituting this value of $\triangle z$ and Equations 4.53 and 4.54 in the plate dynamics for the middle section of the column given by Equation 4.43 leads to

$$
\frac{H}{\triangle z} \frac{dx_j^{(i)}}{dt} = V \frac{\partial y_j^{(i)}}{\partial z} - L \frac{\partial x_j^{(i)}}{\partial z}
\tag{4.55}
$$

These partial differential equations can be converted back to ordinary differential equations using orthogonal collocation. However, in this conversion the order of the differential equations is reduced from $n * (N + 2)$ differential equations with associated equilibrium calculations to $n * (ncol + 2)$ differential equations, where $ncol$ is the number of collocation points. The number of collocation points is much smaller than the number of plates N. This transformation is presented below.

The variables x and y are expressed in terms of Lagrange polynomials, l. We have

$$
x^{(i)}(z, t) = \sum_{k=1}^{ncol+2} l_k(z) x_k^{(i)}(t)
$$

$$
y^{(i)}(z, t) = \sum_{k=1}^{ncol+2} l_k(z) y_k^{(i)}(t)
\tag{4.56}
$$

Therefore, the partial derivatives can be expressed as

$$\frac{\partial x^{(i)}(z,t)}{\partial z}\Big|_{z_j} = \sum_{k=1}^{ncol+2} A_{jk} x_k^{(i)}(t)$$

$$\frac{\partial y^{(i)}(z,t)}{\partial z}\Big|_{z_j} = \sum_{k=1}^{ncol+2} A_{jk} y_k^{(i)}(t) \qquad (4.57)$$

Substituting the above equations in Equation 4.55 results in

$$\frac{H}{\triangle z}\frac{dx_j^{(i)}}{dt} = \sum_{k=1}^{ncol+2} A_{jk}(V y_k^{(i)} - L x_k^{(i)})$$

$$j = 1, 2, \ldots, ncol + 2; \qquad i = 1, 2, \ldots, n \qquad (4.58)$$

The boundary conditions are given by the reboiler and condenser equations below.

$$\frac{dx_1^{(i)}}{dt} = \frac{V}{H_D}(y_1^{(i)} - x_1^{(i)}), \qquad i = 1, 2, \ldots, n \qquad (4.59)$$

$$\frac{dx_B^{(i)}}{dt} = \frac{1}{B}[L(x_{ncol+2}^{(i)} - x_B^{(i)}) - V(y_{ncol+2}^{(i)} - x_B^{(i)})] \qquad (4.60)$$

Similarly, the equilibrium correlations given by Equations 4.50 and 4.51 are converted to

$$y_j^{(i)} = \frac{\alpha_i x_j^{(i)}}{(\sum_{k=1}^n \alpha_k x_j^{(k)})} + \triangle z \sum_{l=1}^{ncol+2} A_{jl} y_l^{(i)}$$

$$i = 1, 2, \ldots, n; j = 1, 2, \ldots, ncol + 2 \qquad (4.61)$$

$$y_{ncol+2}^{(i)} = \frac{\alpha_i x_B^{(i)}}{(\sum_{k=1}^n \alpha_k x_B^{(k)})}; \qquad i = 1, 2, \ldots, n \qquad (4.62)$$

For the partial differential approximation, the matrix $[A]$ corresponds to

$$A_{jk} = \frac{\triangle l_k z}{\triangle z}\Big|_{z\, =\, z_j} \qquad (4.63)$$

If instead of the partial differential approximation which assumes continuity, a discrete approximation is used (Chao and Joseph, 1983a,b; Srivastava and Joseph, 1985, 1987; Stewart et al., 1985) then the matrix $[A]$ becomes:

$$A_{jk} = \frac{l_k z + \delta z - l_k z}{\delta z}\Big|_{z\, =\, z_j} \qquad (4.64)$$

It can be seen that the dynamic representation (in the form of ODEs) of

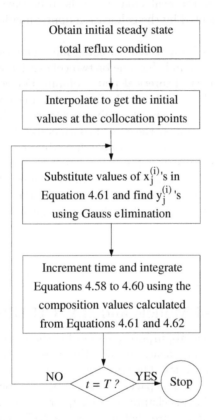

FIGURE 4.16
Flowchart for the Collocation-Based Reduced Order Model for the Constant
Reflux Mode (Reproduced from Diwekar, 1988)

the column (Equations 4.58-4.60) and the vapor-liquid equilibria (Equations 4.44 and 4.45) remain the same. These dynamics equations are integrated numerically to obtain the complete column dynamics for the constant reflux and variable reflux operations. For the variable reflux mode a procedure similar to that used in Chapter 3 is used to vary the reflux ratio at the top. Figure 4.16 illustrates the calculation procedure for the collocation-based reduced order model for simulation of the constant reflux mode.

For columns where the heat balance needs to be taken into account, the above model is augmented by the collocation approximations to the heat balance equations.

This approach has been tested on some ideal and nonideal systems by Diwekar (1988). It was found that one or two collocation points are sufficient to approximate a twenty theoretical plate column. Furthermore, presence of the lighter than light key (LLK) or the heavier than heavy key (HHK) does not create any problems if lower order polynomials are used to approximate the column (Diwekar, 1988), unlike the case of continuous columns (Srivastava and Joseph, 1985, 1987). This is because in batch distillation the LLK disappears very quickly and HHK becomes the heavy key. This property shows promise for this approach especially when other simplified models like the shortcut and modified shortcut models cannot be used to describe the column (for example, highly nonideal systems, systems for which constant molal flow assumptions cannot be used).

4.5   Hierarchy of Models and BATCH-DIST

BATCH-DIST is a general-purpose package for the design, simulation, and optimization of multicomponent, multifraction batch distillation columns operating under different modes. The educational version of this package can be obtained from the author. The package includes simulation models of varying complexity described in this chapter and Chapter 3. The hierarchy of models and guidelines for their use are shown in Figure 4.17. With these models as the basis, the package can accomplish numerous tasks, which are:

- Rapid analysis of column behavior

- Preliminary design

- Optimal design (Chapter 5)

- Evaluating different modes of operation

- Feasible region of operation

- Rapid simulation of large and complex columns (Chapter 7)

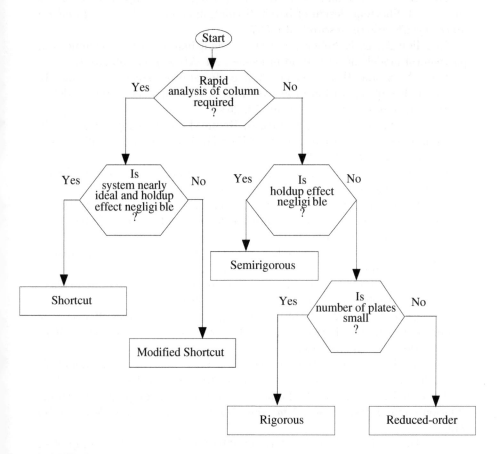

FIGURE 4.17
Guidelines for the Use of the Models (Reproduced from Diwekar and Madhavan, 1991b)

- Control (Chapter 8)

- Rigorous model simulation.

References

Al-Tuwaim M. S. and W. L. Luyben (1991), Multicomponent batch distillation. 3. Shortcut design of batch distillation columns, *Industrial and Engineering Chemistry Research*, **30**, 507.

Benallou A., D. E. Seborg, and D. A. Mellichamp (1986), Dynamic compartmental models of separation processes, *AIChE Journal*, **32**, 1067.

Chao Y. S. and B. Joseph (1983a), Reduced-order steady-state and dynamic models of separation processes: Part I. Development of the model reduction procedure, *AIChE Journal*, **29**, 261.

Chao Y. S. and B. Joseph (1983b), Reduced-order steady-state and dynamic models of separation processes: Part II. Application to nonlinear systems, *AIChE Journal*, **29**, 270.

Chiotti O. J. and O. A. Iribarren (1991), Simplified models for binary batch distillation, *Computers and Chemical Engineering*, **15**, 1.

Diwekar U. M. (1988), *Simulation, design, optimisation, and optimal control of multicomponent batch distillation columns*, Ph.D. Thesis, Indian Institute of Technology, Bombay, India.

Diwekar U. M. (1991), An efficient design method for binary azeotropic batch distillation columns, *AIChE Journal*, **37**, 1571.

Diwekar U. M. (1992a), Unified approach to solving optimal design-control problems in batch distillation, *AIChE Journal*, **38**, 1551.

Diwekar U. M. (1992b), Capturing holdup effects in batch distillation using a lumped parameter short-cut model, Paper presented at the 1992 AIChE Annual Meeting, Miami Beach, FL.

Diwekar U. M. (1994), How simple can it be? — A look at the models for batch distillation, *Computers and Chemical Engineering*, **18** suppl., S451.

Diwekar U. M. and K. P. Madhavan (1991a), Multicomponent batch distillation column design, *Industrial and Engineering Chemistry Research*, **30**, 713.

Diwekar U. M. and K. P. Madhavan (1991b), BATCH-DIST: A comprehensive package for simulation, design, optimization, and optimal control of multicomponent, multifraction batch distillation columns, *Computers and Chemical Engineering*, **15**, 833.

Diwekar U. M., R. K. Malik, and K. P. Madhavan (1987), Optimal reflux rate policy determination for multicomponent batch distillation columns, *Computers and Chemical Engineering*, **11**, 629.

Ellerbe R. W. (1979), *Batch Distillation: Handbook of Separation Techniques for Chemical Engineers*, P. A. Schweitzer, ed., McGraw Hill, New York.

Farhat S., M. Czernicki, L. Pibouleau, and S. Domench (1990), Optimization of multiple-fraction batch distillation by nonlinear programming, *AIChE Journal*, **36**, 1349.

Fenske M. R. (1932), Fractionation of straight run Pennsylvania gasoline, *Industrial Engineering Chemistry*, **24**, 482.

Finlayson B. A. (1972), *The Method of Weighted Residuals and Variational Principles*, Academic Press, NY.

Frazer R. A., W. P. Jones, and S. W. Skan (1937), Approximations to functions and to the solutions of differential equations, Great Britain Aerospace Research Council London. Report and Memo No. 1799. Reprinted in *Great Britain Air Ministry Aerospace Research Communications Technical Report*, **1**, 735.

Galerkin B. G. (1915), Rods and plates. Series in some problems of elastic equilibrium of rods and plates. *Vestn. Inzh. Tech. (USSR)*, **19**, 897.

Gilliland E. R. (1940), Multicomponent rectification. Estimation of the number of theoretical plates as a function of reflux, *Industrial Engineering Chemistry*, **32**, 1220.

Kalagnanam J. R. and U. M. Diwekar (1993), An application of qualitative analysis of ordinary differential equations to azeotropic batch distillation, *AI in Engineering*, **8**, 23.

Lanczos C. (1938), Trigonometric interpolation of empirical and analytic functions, *Journal of Mathematical Physics*, **17**, 123.

Rose A. and L. M. Welshans (1940), Sharpness of separation in batch fractionation, *Industrial Engineering Chemistry*, **32**, 668.

Sørensen E. (1994), *Studies on optimal operation and control of batch distillation columns*, Ph.D. Thesis, University of Trondheim, The Norwegian Institute of Technology, Trondheim, Norway.

Srivastava R. K. and B. Joseph (1984), Simulation of packed-bed separation processes using orthogonal collocation, *Computers and Chemical Engineering*, **8**, 43.

Srivastava R. K. and B. Joseph (1985), Reduced-order models for separation columns - V: Selection of collocation points, *Computers and Chemical Engineering*, **9**, 601.

Srivastava R. K. and B. Joseph (1987), Reduced-order models for staged separation columns - VI: Columns with steep and flat composition profiles, *Computers and Chemical Engineering*, **11**, 165.

Stewart W., K. Levien, and M. Morari (1985), Simulation of fractionation by orthogonal collocation, *Chemical Engineering Science*, **40**, 409.

Underwood A. J. V. (1948), Fractional distillation of multicomponent mixture, *Chemical Engineering Progress*, **44**, 603.

vanDongen D. B. and M. F. Doherty (1985), Design and synthesis of homogeneous azeotropic distillations I. Problem formulation for a single column, *Industrial Engineering Chemistry Fundamentals*, **24**, 454.

Villadsen J. (1970), *Selected Approximation Methods for Chemical Engineering Problems*, Lyngby, Denmark, Instituttet for Kemiteknik.

Villadsen J. and M. L. Michalesen (1978), *Solution of Differential Equation Models by Polynomial Approximation*, Prentice-Hall, Englewood Cliffs, NJ.

Villadsen J. V. and W. E. Stewart (1967), Solution of boundary-value problems by orthogonal collocation, *Chem. Eng. Sci*, **22**, 1483.

Wong K. T. and R. Luus (1980), Model reduction of high-order multistage systems by method of orthogonal collocation, *AIChE Journal*, **29**, 269.

Zuiderweg F. J. (1953), Absatzweise Destillation. Einflub der Bodenzahl, des Rucklaufverhaltnisses und des Holdup auf die Trennscharfe, *Chemie Ingenieur Technik*, **25**, 297.

Exercises

4.1 Derive the analytical formula for $\partial G_c / \partial C_1$ and $\partial G_v / \partial C_1$ for the constant reflux and variable reflux shortcut calculations, respectively.

4.2 A liquid mixture containing 60 mole percent benzene and 40 mole percent toluene is charged in a constant reflux operated batch column. The column has ten plates. Find the composition of the vapor accumulated in the condenser and the liquid remaining in the still when exactly half of the mixture is in the condenser. Assume the relative volatility α_{b-t} is 2.4, and the reflux ratio is 1. Use the shortcut method for calculation.

4.3 An equimolar multicomponent mixture containing meta-, ortho-, and para-mono-nitro-toluene is to be distilled using batch distillation. The column has five theoretical plates. If the distillate composition of the meta-mono-nitro-tolune is to be maintained at 0.90 by varying reflux, is the number of plates sufficient?

4.4 Derive feasibility considerations for the constant reflux design variables N and R and variable reflux design variables $R_{initial}$ and N in terms of the distillate and bottom compositions.

4.5 Provide a flowchart for the simulation of the constant reflux and variable reflux modes of operation using the collocation-based reduced order models.

5

OPTIMIZATION

CONTENTS

The previous chapters concentrated on the design and simulation of batch distillation columns using a hierarchy of models. This chapter presents the design optimization problem. Because batch distillation is a very flexible operation and there are several ways it can operate, it is difficult to analyze all the different possible combinations. Therefore, this chapter focuses only on optimization of the main operating modes of batch distillation. The same analysis, however, can be applied to more complex systems.

The theory of optimization grew out of a problem in the calculus of variation with the *isoperimetric problem* proposed by Queen Dido in 1000 BC. She procured for the founding of Carthage the largest area of land that could be surrounded by the hide of a bull. From the hide she made a rope, which she arranged in a semicircle with the ends against the sea. Queen Dido's intuition was right in finding the solution for the isoperimetric problem, which later proved to be a not-so-simple problem in the calculus of variation. The calculus of variation essentially handles problems where the decision variable is a time-dependent (or integrating-variable-dependent) vector. These problems are also referred to as optimal control problems and are discussed in detail in Chapter 8.

The calculus of variation, or the first systematic theory of optimization, was born on June 4, 1694, when John Bernoulli posed the *brachistochrone* (Greek for "shortest time") problem, and publicly challenged the mathematical world to solve it. The problem is posed as, "what is a slide down which a frictionless object would slip in the least possible time?" Earlier attempts to solve this problem were made by many well-known scientists, including Galileo, who proposed the solution to be a circular arc, an incorrect solution, and Leibnitz, who presented ordinary differential equations without solving them. Then John Bernoulli proved it to be a cycloid. From that point, efforts continued in the area of the calculus of variation, leading to the study of multiple integrals, differential equations, control theory, problem transformation, etc. Although this research was more toward theory and analytical solutions, it developed an abstract setting for the numerical optimization developed after 1945. In fact, in this era of the calculus of variation, the Rayleigh, Ritz, and Galerkin methods of polynomial approximation emerged (also referred to in Chapter 4, Section 4.4 on reduced order collocation models) to solve the two-point boundary value problems encountered in the calculus of variation formulation. However, World War II made scientists aware of numerical optimization and numerical solutions to physics and engineering problems. In 1947 Danzig proposed the simplex algorithm for linear programming problems. Necessary conditions were presented by Kuhn and Tucker in the early 1950s, which formed a focal point for nonlinear programming research. Now numerical optimization techniques constitute a fundamental part of theoretical and practical science and engineering.

The literature on batch distillation optimization problems, surprisingly enough, follows the same trend as optimization theory. The batch distillation optimal reflux policy problem appeared as an example of optimal control theory in many books and papers as early as 1963, when Converse and Gross presented the maximum distillate problem. Also, in general, optimal control problems have received considerable attention in the field of batch distillation. The articles range from a look at different objective functions such as maximum distillate, minimum time, or maximum profit to use of different optimization methods, such as the calculus of variation, maximum principle, dynamic programming, or nonlinear optimization techniques. Very few researchers have looked at the problem of optimizing a specific mode of operation other than the optimal control mode. Due to the abundance of literature on this topic, a separate chapter is devoted to optimal control rather than combining it with optimization in general.

5.1   Objective Functions

Beightler, Phillips, and Wilde (1967) described optimization as a three-step-decision-making process, where the first step is the knowledge of the system and hence is related to the modeling of the process. The previous chapters presented this step for batch distillation. The second step involves finding a measure of system effectiveness. The third step is related to the theory of optimization, which starts with a degrees of freedom analysis and application of a proper optimization algorithm to find the solution. This section deals with the second step of optimization and describes the measure of effectiveness in terms of an objective function and constraints.

In earlier works on batch distillation optimization the objective functions considered were maximizing the sharpness of separation (Houtman and Husain, 1956), maximizing the capacity factor (Luyben, 1971) for constant reflux operation, and maximizing the profit for variable reflux operation (Robinson and Goldman, 1969). Diwekar and co-workers (1989) used profit maximization for the constant reflux and variable reflux modes of operation for single as well as multifraction batch distillation columns.

The sharpness of separation can be attributed approximately to the amount of intermediate fraction distilled as a percentage of total feed. The lower the amount, the sharper the separation. Bowman and Cichelli (1949) presented a very interesting concept of pole height for the binary batch distillation column. A pole height S is defined as the product of the mid-point of the slope of the distillate composition versus the material remaining in the still curve, $dx_D/dB|_{x_D=0.5}$, and the amount of material remaining in the still at that time, $B_{x_D=0.5}$. Figure 5.1 illustrates the concept of pole height. Bowman and Cichelli stated that the pole height is invariant to the initial concentration and provides a good measure for defining the sharpness of separation. Houtman and Husain (1956) used the pole height (S) concept and semi-empirical correlations of pole height with the relative volatility α (for binary separation), the reflux ratio R, the plate holdup H, and the amount remaining in the still B proposed by Zuiderweg (1953), given below, to obtain the optimal value of holdup for maximizing the sharpness of separation.

Zuiderweg equation:

$$E \quad = \quad \alpha^N \tag{5.1}$$

$$\frac{1}{S} \quad = \quad \frac{8}{E}\left[1 + \frac{H}{B_{x_D=0.5}}\left(\frac{E}{4.6\log E} - 1\right)\right]$$

$$+ \quad \frac{1}{cR(\alpha - 1)(1 + \frac{3H}{B_{x_D=0.5}}\log E)} \tag{5.2}$$

where c is an empirical constant.

Houtman and Husain (1956) equated the derivative of Equation 5.2 for S with respect to H to obtain the optimum value of H that gave maximum separation for the minimum reflux ratio.

As stated in Chapter 4, holdup can affect the performance of a batch distillation column in two basic ways: the dynamic flywheel effect, and the steady state capacitance effect. The flywheel effect can be characterized by the parameter $\tau = holdup/(reflux \times distillate\ rate)$. For large τ, the initial composition profile predicted by the zero holdup models departs significantly from the results of a rigorous model. The capacitance effect is observed at the end of the total reflux condition, when the given charge distributes itself throughout the column. Associated with the capacitance effect is the concept of equilibration time. These two effects oppose each other. The objective functions described by Houtman and Husain (1956) and Luyben (1971) considered this factor and presented the concept of optimum holdup. Luyben's objective was to maximize the capacity factor (CF), which was defined as the amount of on-specification products produced per time, i.e., the sum of the distillate

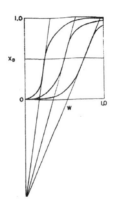

FIGURE 5.1

The Pole Height Concept (Reproduced from Bowman and Cichelli, 1949)

D and the bottoms B:

$$CF = \frac{B + D}{T_{eq} + T} \qquad (5.3)$$

where T_{eq} is the equilibration time and T is the batch time. He considered the effect of condenser holdup and plate holdup for different binary systems.

Robinson and Goldman (1969) used a profit function for optimizing variable reflux operation. However, the form of their profit function is not reported in the literature. Diwekar et al. (1989) used the profit function P given below.

$$P = \text{annual sales value of distillate} - \text{annual cost of distillation} \qquad (5.4)$$

Their annual cost of distillation was based on the cost function of the continuous column given by Happel (1958), which involved the following items:

- the amortized cost of distillation equipment per year: $\frac{c_1 V N}{G_a}$

- the amortized cost of the reboiler and condenser per year: $\frac{c_2 V}{G_b}$

- the annual cost of steam and coolant: $\frac{24(365)c_3 V T}{T + t_s}$

where

c_1 is the amortized incremental investment cost [$/m^2/plate/year],

c_2 is the amortized incremental cost of the equipment [$/m^2/year],

c_3 is the cost of the steam and coolant to vaporize or condense, respectively, 1 kg of distillate [$/kg]

G_a is the allowable vapor velocity [kg/hr/m^2]

G_b is the vapor handling capacity of the equipment [kg/hr/m^2]

N is the number of plates

T is the batch time [hr]

t_s is the setup time for each batch [hr]

V is the vapor boilup rate [kg/hr]

Assuming D = number of moles of product distilled, P_r = sales value of the product, $/mole, and C_0 = cost of feed, $/mole, the profit function can be written as:

$$P = \frac{24(365)DP_r}{T + t_s} - \frac{c_1 V N}{G_a} - \frac{c_2 V}{G_b} - \frac{24(365)c_3 V T}{T + t_s} \qquad (5.5)$$

Although earlier researchers like Houtman and Husain (1956), Robinson and Goldman (1969), and Luyben (1971) presented design optimization problems in batch distillation, the solutions were presented in terms of single variable optimization through graphs. Diwekar and co-workers presented the complete analysis (all three steps) and used numerical multivariable optimization techniques to obtain the optimal solution. Therefore, the third decision-making step in optimization, described below, is based on Diwekar (1988) and Diwekar et al. (1989).

5.2   Degrees of Freedom Analysis

The degrees of freedom analysis provides the number of decision variables one can change to obtain the optimum design and is crucial in numerical optimization. It is easier to understand the degrees of freedom analysis of batch distillation in the context of the shortcut method. Therefore, the analysis is presented below using the shortcut method equations and is derived from Diwekar (1988) and Diwekar et al. (1989). Since the shortcut method is qualitatively similar to the higher order models, the same analysis applies to the rigorous model, although holdup effects are neglected in the shortcut model.

We have seen that some of the earlier studies in the optimization of batch columns used holdup as a decision variable (Houtman and Husain, 1956; Luyben, 1971). However, it should be noted that holdup is mainly a function of tray hydraulics, and in almost all of the models of batch distillation, the effects of tray hydraulics are not considered (Tomazi and Waggoner, 1993, presented a preliminary analysis of tray hydraulics in the context of variable reflux operation). Furthermore, depending on the tray design, the holdup may be a discrete variable, making the problem more difficult to handle than problems presented in earlier works. Nevertheless, it is an important factor. However, for the first-cut analysis presented here, the holdup could be assumed to be specified for the column and hence can be omitted from the degrees of freedom analysis. Also, this analysis is presented for the three modes of operation for a single fraction condition.

The versatility of batch distillation allows for a large number of configurations which can lead to many more degrees of freedom than what is presented below.

Constant Reflux Mode

To make the unsteady state process more understandable, the degrees of freedom analysis is presented in terms of t discrete integration steps. Table 5.1 presents the degrees of freedom analysis for the constant reflux mode based on the shortcut method. As shown in Table 5.1, for the constant reflux mode, the unknowns include reflux ratio R, number of plates N, vapor boilup rate V, bottom composition at each time step (except at $t = 0$), and distillate composition at each time step. The differential material balance equation (integrated from the initial bottom composition equal to the feed composition) which provides the relation between the change in bottom composition of the key component 1 $x_B^{(1)}$ (integration variable for the time-implicit shortcut method equations) and the bottom composition of all other components $x_B^{(i)}$, $i = 2, 3, \ldots, n$. Once the bottom composition is known, the Hengestebeck–Geddes equation relates the distillate composition of components other than the key component $(x_D^{(i)}, i = 2, 3, \ldots, n)$ to the bottom composition and the key component distillate composition. However, this equation introduces an additional unknown, the constant C_1 which can be calculated using the fact that the sum of distillate composition of all the components is equal to unity. The equation for total time calculation and the implicit relation derived from the FUG equations eliminate an additional two degrees of freedom, leading to a total of three degrees of freedom. Diwekar et al. (1989) selected the number of plates N, the reflux ratio R, and the vapor boilup rate V as the decision variables.

It should be noted that the vapor boilup rate is dependent on the flooding and weeping conditions of the column was not considered in the early analysis presented by Diwekar et al. (1989).

Variable Reflux Mode

The degrees of freedom analysis for the variable reflux mode is presented in Table 5.2. Similar to the constant reflux mode, the unknowns include number of plates N, vapor boilup rate V, bottom composition at each time step except initial, and distillate composition of components other than the key component for each time step. However, in this case the reflux ratio has to be evaluated at each time step instead of the distillate composition of the key component. Again following the shortcut method, the differential material balance can be used to evaluate the bottom composition at each time step. Distillate composition for all components can be obtained from the Hengestebeck–Geddes equation along with the equation for C_1 calculation. The equation for time derived from Equation 2.16 in Chapter 2 and the FUG equations reduce the total degrees of freedom to 3 which is same as the constant reflux mode. Diwekar and co-workers (Diwekar et al., 1989) selected the initial reflux ratio R_0, the stopping criterion in terms of the final still composition $x_{B\infty}$, and the vapor boilup rate V as the decision variables. Since the number of plates is implicitly related to the initial reflux ratio, it could not be used as another decision variable as in the constant reflux mode. In the case of constant reflux operation, the average distillate composition decides the stopping criterion. However, in this case the distillate composition is always maintained constant, so the final stopping criterion has to be imposed. Therefore, the second decision variable is chosen to be the final stopping criterion.

Optimal Reflux Mode

The optimal reflux mode of operation offers much more flexibility than the other two modes above and hence the analysis leads to an increased number of degrees of freedom, as shown in Table 5.3. It can be easily seen that for the optimal reflux mode of operation we have an additional degree of freedom at each integration step, which is attributed to the calculation of the reflux ratio at each step. Logsdon et al. (1990) included the number of plates N, the stopping criterion T, and vapor boilup rate V as the decision variables along with the optimal reflux profile.

5.3   Feasibility Considerations

The degrees of freedom analysis presented in the earlier section resulted in a maximum of three degrees of freedom for the single fraction case (in the

TABLE 5.1
Degrees of Freedom Analysis for the Constant Reflux Mode

Unknowns	Number
Reflux Ratio, R	1
Number of Plates, N	1
Vapor Boilup Rate, V	1
Bottom Composition at Each Time Step (except at $t = 0$), $x_B^{(i)}$, $i = 1, 2, \ldots, n$	$n \times (t-1)$
Distillate Composition at Each Time Step, $x_D^{(i)}$, $i = 1, 2, \ldots, n$	$n \times t$
Integration Termination Criterion	1
Total	$2n \times t - n + 4$

Equations	Number
Differential Material Balance Equations at Each Time Step Except Initial $$x_{B_{new}}^{(i)} = x_{B_{old}}^{(i)} + \frac{\triangle x_B^{(1)} \left(x_D^{(i)} - x_B^{(i)} \right)_{old}}{\left(x_D^{(1)} - x_B^{(1)} \right)_{old}}, i = 1, 2, \ldots, n$$	$n \times (t-1)$
Hengestebeck–Geddes Equation at Each Time Step $$x_D^{(i)} = \left(\frac{\alpha_i}{\alpha_1} \right)^{C_1} \frac{x_D^{(1)}}{x_B^{(1)}} x_B^{(i)}, i = 2, 3, \ldots, n$$	$(n-1) \times t$
Summation of Distillate Composition (C_1 Calculation) $$\sum_{i=1}^{n} \left[\left(\frac{\alpha_i}{\alpha_1} \right)^{C_1} \frac{x_D^{(1)}}{x_B^{(1)}} x_B^{(i)} \right] = 1$$	
Equation for Time (from Equations 2.1 and 2.12) $$T = \frac{F(R+1)}{V}[1 - \exp(\sum_{x_{B\infty}^{(1)}}^{x_F^{(1)}} \frac{\triangle x_B^{(1)}}{x_D^{(1)} - x_B^{(1)}})]$$	1
Implicit Relation (FUG Equations) $$x_D^{(1)} = f(R, N, x_B(i))$$	t
Total	$2n \times t - n + 1$

Degrees of Freedom	
$2n \times t - n + 4 - (2n \times t - n + 1) =$	3

TABLE 5.2
Degrees of Freedom Analysis for the Variable Reflux Mode

Unknowns	Number
Reflux Ratio, R	t
Number of Plates, N	1
Vapor Boilup Rate, V	1
Bottom Composition at Each Time Step (except at $t = 0$), $x_B^{(i)}, \quad i = 1, 2, \ldots, n$	$n \times (t - 1)$
Distillate Composition at Each Time Step $x_D^{(i)}, \quad i = 2, \ldots, n$	$(n - 1) \times t$
Distillate Composition of the Key Component (which remains fixed throughout), $x_D^{(1)}$	1
Integration Termination Criterion	1
Total	$2n \times t - n + 4$

Equations	Number
Differential Material Balance Equations at Each Time Step Except Initial $x_{B_{new}}^{(i)} = x_{B_{old}}^{(i)} + \frac{\Delta x_B^{(1)}\left(x_D^{(i)} - x_B^{(i)}\right)}{\left(x_D^{(1)} - x_B^{(1)}\right)_{old}}, i = 1, 2, \ldots, n$	$n \times (t - 1)$
Hengestebeck–Geddes Equation at Each Time Step $x_D^{(i)} = \left(\frac{\alpha_i}{\alpha_1}\right)^{C_1} \frac{x_D^{(1)}}{x_B^{(1)}} x_B^{(i)}, i = 2, 3, \ldots, n$	$(n - 1) \times t$
Summation of Distillate Composition (C_1 Calculation) $\sum_{i=1}^{n}\left[\left(\frac{\alpha_i}{\alpha_1}\right)^{C_1} \frac{x_D^{(1)}}{x_B^{(1)}} x_B^{(i)}\right] = 1$	
Equation for Time (from Equation 2.16) $T = \frac{F(x_D^{(1)} - x_F^{(1)})}{V}[1 - \sum_{x_{B\infty}^{(1)}}^{x_F^{(1)}} \frac{\Delta x_B^{(1)}(R+1)}{(x_D^{(1)} - x_B^{(1)})^2}]$	1
Implicit Relation (FUG Equations) $R = f(x_D^{(1)}, N, x_B(i))$	t
Total	$2n \times t - n + 1$

Degrees of Freedom	
$2n \times t - n + 4 - (2n \times t - n + 1) =$	3

TABLE 5.3

Degrees of Freedom Analysis for the Optimal Reflux Mode

Unknowns	Number
Reflux Ratio, R	t
Number of Plates, N	1
Vapor Boilup Rate, V	1
Bottom Composition at Each Time Step (except at $t=0$), $x_B^{(i)}$, $i=1,2,\ldots,n$	$n \times (t-1)$
Distillate Composition at Each Time Step $x_D^{(i)}$, $i=1,2,\ldots,n$	$n \times t$
Purity Consideration x_{Dav}	1
Integration Termination Criterion	1
Total	$2n \times t - n + t + 4$

Equations	Number
Differential Material Balance Equations at Each Time Step Except Initial $x_{B_{new}}^{(i)} = x_{B_{old}}^{(i)} + \frac{\Delta x_B^{(1)}\left(x_D^{(i)} - x_B^{(i)}\right)_{old}}{\left(x_D^{(1)} - x_B^{(1)}\right)_{old}}, i = 1,2,\ldots,n$	$n \times (t-1)$
Hengestebeck–Geddes Equation at Each Time Step $x_D^{(i)} = \left(\frac{\alpha_i}{\alpha_1}\right)^{C_1} \frac{x_D^{(1)}}{x_B^{(1)}} x_B^{(i)}, i = 2,3,\ldots,n$	$(n-1) \times t$
Summation of Distillate Composition (C_1 Calculation) $\sum_{i=1}^{n}\left[\left(\frac{\alpha_i}{\alpha_1}\right)^{C_1} \frac{x_D^{(1)}}{x_B^{(1)}} x_B^{(i)}\right] = 1$	
Equation for Time (from Equations 2.1 and 2.12) $T = \frac{F(R+1)}{V}[1 - exp(\sum_{x_{B\infty}^{(1)}}^{x_F^{(1)}} \frac{\Delta x_B^{(1)}}{x_D^{(1)} - x_B^{(1)}})]$	1
Implicit Relation (FUG Equations) $x_D^{(1)} = f(R, N, x_B(i))$	t
Total	$2n \times t - n + 1$

Degrees of Freedom	
$2n \times t - n + t + 4 - (2n \times t - n + 1) =$	$t + 3$

absence of holdup) for both the variable reflux and the constant reflux operating conditions. It was also shown that the same three degrees of freedom are available for the optimal reflux problem. The optimal decision variables could be the initial reflux ratio R_0 (fixed by the purity constraint), the number of plates N, the stopping criterion T, and the vapor boil-up rate V. In order to maintain the feasibility of design, appropriate constraints on the variables are to be imposed, especially for the design variables like number of plates N and the reflux ratio R. The shortcut method helps to identify these bounds on the design parameters. The bounds on the parameters depend on the operating modes. For the variable reflux, the constant reflux, and the optimal reflux modes, the feasible region of operation has been identified using the shortcut method (Diwekar et al., 1989; Diwekar, 1992). The following paragraph describes these feasibility considerations, which are applicable to the zero holdup models and can be extended to holdup effects with some modifications.

For the variable reflux case, the value of $x_D^{(1)}$ $(=\ x_D^*)$ (the key component is 1) is specified, and the reflux ratio is at its minimum value at the initial conditions. Therefore, the lower bound to R_0 is the value of R_{min} at the initial conditions (which will need an infinite number of plates for the given separation $x_D^{(1)}\ =\ x_D^*$). The upper limit is governed by N_{min} at the termination criterion (total reflux condition). To obtain the upper limit on R, N_{min} is calculated at the terminal condition and is taken as the limiting value of N. The value of R corresponding to this N in order to attain the distillate purity specified by x_D^* at the initial condition is the value of R_{max}, the upper bound to R_0. The concept of R_{max} is shown in Figure 5.2. Similarly, if the termination criterion is expressed in terms of the still composition of the more volatile component, then the still composition should be less than the feed composition or the distillate composition, and should be greater than or equal to zero. As stated earlier, the vapor boilup rate is bounded by the flooding and weeping conditions.

For the constant reflux and the optimal reflux conditions, the initial value of R has a lower bound, defined by the specified average distillate composition of the key component x_D^*. The initial distillate composition of the key (more volatile) component is highest at the beginning and decreases as distillation progresses. If the initial value of R is such that the distillate composition of the key component is less than the specified average, then the goal of attaining the specified average purity is impossible to meet for the given number of plates. This criterion provides the lower limit R_{MIN} on the initial R, where R_{MIN} is defined as the value of R required to obtain the distillate composition of the key component $(x_D^{(1)})$ equal to the specified average distillate composition (x_D^*) at the initial conditions for the given N. The concept is illustrated in Figure 5.3. Similarly, the value of N is also restricted by the lower bound N_{min} using the same criterion. Here N_{min} is the Fenske value of the minimum number of plates for the specified average distillate composition. All the bounds for the three operating modes based on the shortcut method are presented in tabular form in Table 5.4.

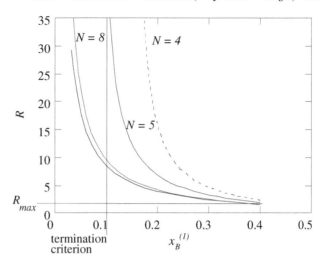

FIGURE 5.2
Concept of R_{max}

5.4   Problem Solution

The batch distillation design optimization problems described in this chapter involve scalar decisions to be made, subject to the model equations and constraints related to the product specifications. As the model equations are complicated, one has to resort to numerical integration techniques which essentially involve solutions of nonlinear algebraic equations at each time step. The nonlinear nature of objective function (e.g., profit, capacity), model equations, and other constraints calls for nonlinear programming optimization techniques.

In general, a nonlinear programming problem (NLP) can be represented as follows:

$$\text{Optimize}_{x} \quad Z \;=\; z(x) \tag{5.6}$$

subject to

$$h(x) \;=\; 0 \tag{5.7}$$

$$g(x) \;\leq\; 0 \tag{5.8}$$

The goal of an optimization problem is to determine the decision variables x that optimize some aspect of the model represented by the objective function Z, while ensuring that the model operates within established limits enforced by the equality constraints h and inequality constraints g. The above

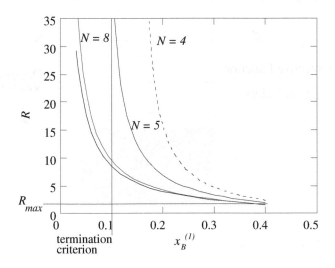

FIGURE 5.3
Concept of R_{MIN}

TABLE 5.4
Feasible Region for Multicomponent Batch Distillation Columns

Variable Reflux	Constant Reflux	Optimal Reflux
Final Still Composition		
$0 \leq x_{B_\infty}^{(1)} \leq x_D^{(1)}$		
Distillate Composition		
$x_B^{(1)} \leq x_D^{(1)} \leq 1$		
Reflux Ratio		
$1 \leq \dfrac{R_{initial}}{R_{min}} \leq \dfrac{R_{max}}{R_{min}}$	$1 \leq \dfrac{R}{R_{MIN}} \leq \infty$	
Number of Plates		
$N_{minf} \leq N$	$N_{min} \leq N$	

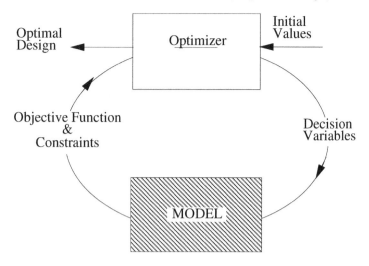

FIGURE 5.4
Pictorial Representation of the NLP Optimization Framework

formulation represents an NLP problem, for which a generalized iterative solution procedure is illustrated schematically in Figure 5.4. As seen in the figure, the optimizer invokes the model with a set of values of decision variables x. The model simulates the phenomena and calculates the objective function and constraints. This information is utilized by the optimizer to calculate a new set of decision variables. This iterative sequence is continued until the optimization criteria pertaining to the optimization algorithm are satisfied. The batch distillation design optimization problem easily can be fit to this description; the model equations and the bounds are part of the equality/inequality constraints and bounds, and the objective such as profit, capacity, etc. can be calculated as a function of decision variables.

Recent advances in constrained nonlinear optimization techniques provide better choices for solving large-scale NLP problems. The most popular of these methods are generalized reduced gradient (GRG) and sequential quadratic programming (SQP) methods and their variants. Among generalized reduced gradient methods, the most widely used algorithms are GRG2 and MINOS (Gill et al., 1981). Most literature on large-scale chemical flowsheet optimization favors the SQP method. In the SQP technique, at each iteration step the problem is approximated as a quadratic program where the objective function is quadratic and the constraints are linear. Similar to linear programming, the special features of a quadratic objective function are exploited to solve the problem more efficiently. The quadratic programming subproblem is solved at each step to obtain the next trial point. This procedure is repeated until the optimum is reached.

There are a large number of codes available for NLP optimization, including MINOS, CONOPT (GRG based), NPSOL (SQP-based), GAMS, and FSQP. Also, many mathematical libraries, such as NAG, IMSL, and HAR-WELL, have different optimization codes embedded in them. However, a discussion of all the different features of NLP methods and a listing of all the accessible code names is beyond the scope of this book. For more details, readers can refer to the *Optimization Software Guide* by More and Wright (1993) from SIAM books.

Chapter 1 presented a few examples of the different kinds of batch distillation columns. Combined with different operating modes, the possible combinations for a single column tend to be very high. There is also an option of using a sequence of columns instead of a single column. If the optimization problem also involves decisions such as number of kind of columns, type of operating mode, etc., along with finding optimal values of the design variables such as reflux ratio, vapor boilup rate, number of plates, etc., this leads to a very challenging area of synthesis problems.

The NLP optimization described above cannot be readily applicable to such problems. Synthesis methods are well established in the context of continuous processes and excellent reviews are available on this subject (Stephanopoulos, 1980; Nishida et al., 1981; Grossmann, 1990). The state of the art techniques used in the solution of synthesis problems include: a) the heuristic approach, which relies on intuition and engineering knowledge, b) the physical insight approach, which is based on exploiting basic physical principles, and c) the optimization approach, which uses mathematical programming techniques (Grossmann, 1990). For batch distillation, vanDongen and Doherty (1985a, b) and Bernot et al. (1990, 1991) have presented a synthesis approach for ternary and quaternary azeotropic systems based on physical insights and heuristics. Their approach is discussed in Chapter 7. The optimization approach to synthesis involves 1) formulation of a superstructure incorporating all the alternative configurations and 2) modeling the superstructure as a mixed integer nonlinear programming (MINLP) problem of the form

$$\text{Optimize} \quad Z = z(x, y) \tag{5.9}$$
$$x, y$$

subject to

$$h(x, y) = 0 \tag{5.10}$$

$$g(x, y) \leq 0 \tag{5.11}$$

where the continuous variable x represents design variables such as vapor boilup rate, reflux ratio and the binary variables y denote the potential existence ($y = 1$) or absence ($y = 0$) of configurations.

Algorithms for solving MINLP problems include the branch and bound method (Beale, 1977; Gupta, 1980), the generalized Benders decomposition (GBD) method (Benders, 1962; Geoffrion, 1972), and the outer approximation

(OA) method (Duran and Grossmann, 1986). The GBD and OA algorithms are, in general, more efficient than the branch and bound method. These algorithms for solving the above MINLP problem consist of solving at each major iteration an NLP subproblem (with all 0-1 variables fixed) and a mixed integer linear programming (MILP) master problem, as shown in Figure 5.5. The NLP subproblem has the role of optimizing the continuous variables and providing the upper bound to the MINLP solution. The MILP master problem predicts the lower bound to the MINLP as well as updates the values of the 0-1 variables for each major iteration. The predicted lower bounds increase monotonically as the cycle of major iterations proceeds, and the search is terminated when the predicted lower bound coincides or exceeds the current upper bound.

The main difference between GBD and OA is in the definition of the master problem. In general, the OA algorithm requires fewer iterations than GBD but involves the solution of a larger master problem. Variants of these algorithms include the outer approximation/equality relaxation (OA/ER) strategy of Kocis and Grossmann (1987), extensions of GBD using a partitioning strategy (Floudas et al., 1989), the augmented penalty function/OA/ER algorithm (OA/ER/AP) of Viswanathan and Grossmann (1990), and the GBD/OA/ER/AP algorithm of Diwekar et al. (1992).

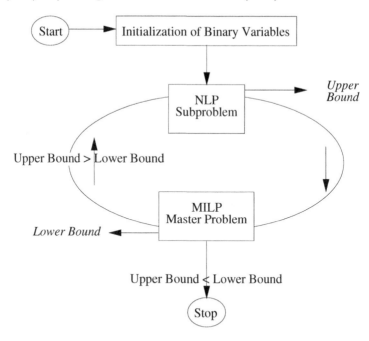

FIGURE 5.5
Main Steps in GBD and OA Algorithms for MINLP

In the area of batch distillation synthesis, although Sundaram and Evans

(1993) identified the associated difficulties and presented a preliminary MINLP formulation, they solved only the underlying optimal control problem as an NLP, similar to the problem presented by Farhat et al. (1991) (please refer to Chapter 8 for details of this approach). The batch distillation synthesis problem as an MINLP optimization problem thus remains open for researchers at this stage.

References

Beale E. M. (1977), *Integer Programming: The State of the Art in Numerical Analysis*, Academic Press, London, United Kingdom.

Benders J. F. (1962), Partitioning for solving mixed-variables programming problems, *Numerical Mathematics*, **4**, 238.

Beightler C. S., D. T. Phillips, and D. J. Wilde (1967),*Foundations of Optimization*, Prentice-Hall Inc., Englewood Cliffs, NJ.

Bernot C., M. F. Doherty, and M. F. Malone (1990), Patterns of composition changes in multicomponent batch distillation, *Chemical Engineering Science*, **45**, 1207.

Bernot C., M. F. Doherty, and M. F. Malone (1991), Feasibility and separation sequencing in multicomponent batch distillation, *Chemical Engineering Science*, **46**, 1311.

Bowman J. R. and M. T. Cichelli (1949), Batch distillation: minimum number of plates and minimum reflux, *Industrial and Engineering Chemistry*, **41**, 1985.

Converse A. O. and G. D. Gross (1963), Optimal distillate policy in batch distillation, *Industrial Engineering Chemistry Fundamentals*, **2**, 217.

Diwekar U. M. (1988), *Simulation, design, optimisation, and optimal control of multicomponent batch distillation columns*, Ph.D. Thesis, Indian Institute of Technology, Bombay, India.

Diwekar U. M. (1992), Unified approach to solving optimal design-control problems in batch distillation, *AIChE Journal*, **38**, 1551.

Diwekar U. M., I. E. Grossmann, and E. S. Rubin (1992), An MINLP process synthesizer for a sequential modular simulator, *Industrial and Engineering Chemistry Research*, **31**, 313.

Diwekar U. M., K. P. Madhavan, and R. E. Swaney (1989), Optimization of multicomponent batch distillation column, *Industrial and Engineering Chemistry Research*, **28**, 1011.

Duran M. A. and I. E. Grossmann (1986), An outer-approximation algorithm for a class of Mixed-Integer Nonlinear Programs, *Mathematical Programming*, **36**, 307.

Farhat S., M. Czernicki, L. Pibouleau, and S. Domench (1990), Optimization of multiple-fraction batch distillation by nonlinear programming, *AIChE Journal*, **36**, 1349.

Floudas C. S., A. Aggrawal, and A. R. Ciric (1989), A global optimum search for nonconvex NLP and MINLP problems, *Computers and Chemical Engineering*, **13**, 1117.

Geoffrion A. M. (1972), Generalized Benders decomposition, *Journal of Optimization Theory and Applications*, **10**, 237.

Gill P. E., W. Murray, and M. H. Wright (1981), *Practical Optimization*, Academic Press, London, United Kingdom.

Grossmann I. E. (1990), MINLP optimization strategies and algorithms for process synthesis, *Proceedings of Foundations of Computer Aided Process Design 89*, Siirola et al. Eds., Elsevier, Amsterdam, The Netherlands.

Gupta O. K. (1980), *Branch and bound experiments in nonlinear integer programming*, Ph.D, Thesis, Purdue University, West Lafayette, Indiana, USA.

Happel J. (1958), *Chemical Process Economics*, John Wiley & Sons Inc., NY.

Houtman J. P. W. and A. Husain (1956), Design calculations for batch distillation column, *Chemical Engineering Science*, **5**, 178.

Kocis G. R. and I. E. Grossmann (1987), Relaxation strategy for the structural optimization of process flowsheets, *Industrial and Engineering Chemistry Research*, **26**, 1869.

Logsdon J. S., U. M. Diwekar, and L. T. Biegler (1990), On the simultaneous optimal design and operation of batch distillation columns, *Chemical Engineering Research and Design*, **68**, 434.

Luyben W. L. (1971), Some practical aspects of optimal batch distillation design, *Industrial and Engineering Chemistry Process Design, Development*, **10**, 54.

More J. J. and S. Wright (1993), *Optimization Software Guide*, Society for Industrial and Applied Mathematics, Philadelphia, PA.

Nishida N., G. Stephanopoulos, and A. W. Westerberg (1981), A review of process synthesis, *AIChE Journal*, **27**, 321.

Robinson E. R. and M. R. Goldman (1969), The simulation of multicomponent batch distillation processes on a small digital computer, *British Chemical Engineer*, **14**, 809.

Stephanopoulos G. (1980), Synthesis of process flowsheets: An adventure in heuristic design or a utopia of mathematical programming, *Proceedings of International Conference on Foundations of Computer Aided Chemical Process Design*, **2**, 439.

Sundaram S. and Evans L.B. (1993), Synthesis of separations by batch distillation, *Ind. Eng. Chem. Res.*, **32**, 500.

Tomazi K. G. and R. C. Waggoner (1993), Batch distillation strategies constrained by tray hydraulics effects, Paper presented at the AIChE Spring National Meeting, Houston, TX.

vanDongen D. B. and M. F. Doherty (1985a), Design and synthesis of homogeneous azeotropic distillations - I: Problem formulation for single column, *Industrial Engineering Chemistry Fundamentals*, **24**, 454.

vanDongen D. B. and M. F. Doherty (1985b), On the dynamics of distillation processes VI: Batch distillation, *Chemical Engineering Science*, **40**, 2087.

Viswanathan J. and I. E. Grossmann (1990), A combined penalty function and outer-approximation method for MINLP optimization, *Computers and Chemical Engineering*, **14**, 769.

Zuiderweg F. J. (1953), Absatzweise Destillation. Einflub der Bodenzahl, des Rucklaufverhaltnisses und des Holdup auf die Trennscharfe, *Chemie Ingenieur Technik*, **25**, 297.

Exercises

5.1 Present the degrees of freedom analysis for the maximum distillate, minimum time, and maximum profit problems described in Chapter 8.

5.2 Present the degrees of freedom analysis using the rigorous dynamic model for batch distillation and show that if holdup is kept constant (which is a function of tray design) then the analysis results in the same degrees of freedom as that of the shortcut method presented in this chapter.

6

COMPLEX COLUMNS

CONTENTS

The transient nature of batch distillation allows for configuring the column in a number of different ways, a few examples of which were shown in Chapter 1. The column at the right in Figure 6.1 is a conventional batch distillation column with a reboiler at the bottom and a condenser at the top, which essentially performs the rectifying operation. In contrast to a conventional batch distillation column, the inverted column (Figure 6.1, left) has its storage vessel at the top and the products leave the column at the bottom. Thus mixtures with a small amount of the light component can be separated by removing the heavy component as the bottom product. These two columns are comparable to the rectifying and stripping parts of a continuous distillation column but with additional flexibility.

Devidyan et al. (1994) presented a batch distillation column that combines both the rectifying and stripping sections, the middle vessel column (Figure 6.2, left). Although this column has not been investigated completely, preliminary analysis has demonstrated that it provides high flexibility and that it is able to remove both light and heavy impurities by having three product vessels. For example, the composition of the most volatile component in a rectifier decreases with time for the constant reflux operation and the composition of the least volatile component in a stripper also decreases with time for the constant reboil operation. In the middle vessel column, however, these effects can be reversed by setting the vapor ratios for the top and the bottom parts of the column appropriately.

Skogestad et al. (1995, 1997) described a new column called a multivessel column (Figure 6.2, right) and showed that the column can obtain purer products at the end of a total reflux operation. With this column it is possible to separate more than three components at a time by installing enough

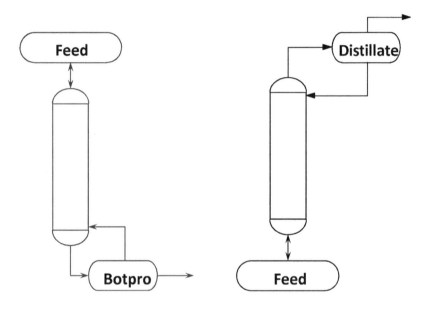

FIGURE 6.1
Inverted Batch Distillation Column (Stripper) and Conventional Batch Distillation Column (Rectifier)

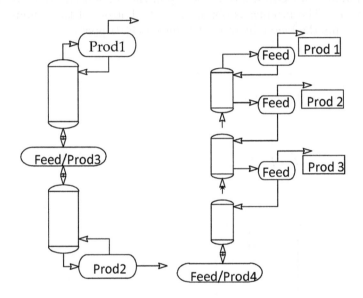

FIGURE 6.2
Middle Vessel Batch Distillation Column and Multivessel Batch Distillation
Column

intermediate vessels. Its design, however, is less flexible than the design of a conventional batch distillation column. The following sections describe these different column configurations in more detail.

6.1   Inverted Column

Figure 6.3 presents the schematic of the inverted column or batch stripper. In this column, the feed tank is at the top of the column and the reboiler is at the bottom. The product is obtained at the bottom of the column. This section presents the hierarchy of models for batch strippers.

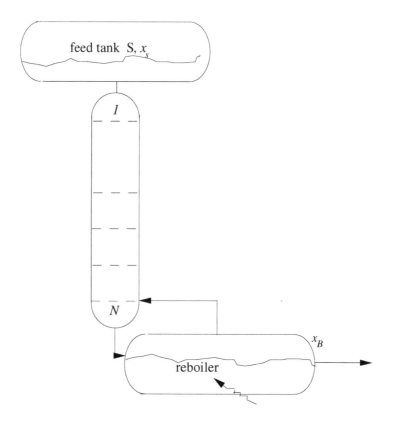

FIGURE 6.3
Batch Stripper

6.1.1   Shortcut Method

This section presents a shortcut model for the inverted batch distillation column (stripper), based on the shortcut model for the batch rectifier (Diwekar and Madhavan, 1991). This model was developed by Lotter and Diwekar (1997).

Similar to the shortcut method for the batch distillation rectifier, the main assumption of the shortcut method for the stripper is that the batch distillation column is regarded as a continuous column with changing feed composition at each time step. Other assumptions include constant relative volatility (ideal systems), equal molal overflow, and negligible plate holdups. Thermodynamic equilibrium is assumed between the vapor and liquid leaving each plate (theoretical plate) and the composition of the liquid leaving the still is assumed to be the same as that of the still (well mixed reservoir). Although the basis of the shortcut method for the stripper is similar to the shortcut method for the rectifier, i.e, based on FUG (Fenske-Underwood-Gilliland) equations, there are differences in the column behaviors. The first difference is the definition of the reflux and reboil ratio. Unlike the reflux ratio, the reboil ratio does not vary between zero and infinity but between one and infinity. This is due to the definition of the reboil ratio (Equation 6.1). This difference also is reflected in various equations for the shortcut procedure.

Reboil ratio and reflux ratio definitions

$$\text{Reflux ratio, } R = \frac{L}{dD/dt}$$

$$\text{Reboil ratio, } Rb = \frac{V_B}{dB/dt} \tag{6.1}$$

We assume that the distillation is carried out at constant boilup rate V_B. The constant molal overflow assumption leads to the following overall material balance equation.

$$\frac{dS}{dt} = -\frac{V_B}{Rb}, \quad S_0 = F \tag{6.2}$$

where F is the feed to the still, S is the amount remaining in the still, and Rb is the reboil ratio. The still composition and bottom composition are given by x_s and x_B, respectively.

$$x_B^{(k)}\frac{V_B}{Rb} = -\frac{d(x_s^{(k)}S)}{dt} \tag{6.3}$$

$$\frac{dx_s^{(k)}}{dt} = \frac{V_B}{(Rb)S}(x_s^{(k)} - x_B^{(k)}), \quad x_{s0}^{(k)} = x_F^{(k)} \tag{6.4}$$

Equation 6.4 relates the change in the still composition to the bottom product composition. The above two mass balance equations can be converted to time-implicit equations similar to the Rayleigh equation for simple distillation and for the batch rectifier. We will call this equation the modified Rayleigh equation. The following is the finite difference approximation to this modified Rayleigh equation for n components expressed in terms of the key component k.

$$x_{s_{new}}^{(i)} = x_{s_{old}}^{(i)} + \frac{\triangle x_s^{(k)} \left(x_B^{(i)} - x_s^{(i)} \right)_{old}}{\left(x_B^{(k)} - x_s^{(k)} \right)_{old}}, \qquad i = 1, 2, \ldots, n \qquad (6.5)$$

As seen in the case of the rectifier, the functional relationship between the end compositions is crucial for the simulation of the complete operation and it is here that we need to use the design equations described by the modified FUG method.

Functional Relationship Between x_B and x_s:

At each instant, there is a change in the still composition of the key component, resulting in changes in the still composition of all the other components calculated by the differential material balance equations (Equation 6.5). For the rectifier, the Hengestebeck–Geddes relation provides the relation between distillate composition and bottom composition. For the stripper, this equation translates into the following equation.

The Hengestebeck–Geddes equation for the stripper.

$$x_B^{(i)} = \left(\frac{\alpha_i}{\alpha_k} \right)^{-C_B} \frac{x_s^{(i)}}{x_s^{(k)}} x_B^{(k)}, i = 1, 2, \ldots, n(i \neq k) \qquad (6.6)$$

where C_B is the Hengestebeck–Geddes (HG) constant for the bottom section of the column (stripper). Fenske derived an equation to calculate the minimum number of plates for a continuous distillation column with constant relative volatilities in terms of the distillate composition and feed composition. It can be easily shown that the Fenske equation for a stripper is given by:

The Fenske equation:

$$Nb_{min} = \frac{\ln\left[\frac{x_s^{(i)}}{x_s^{(k)}} \frac{x_B^{(k)}}{x_B^{(i)}} \right]}{\ln\left[\frac{\alpha_i}{\alpha_k} \right]} \qquad (6.7)$$

The minimum number of plates is the number of equilibrium plates required for a separation at total reflux conditions and is thus a boundary of the operating conditions. If the Hengestebeck–Geddes equation is compared with the Fenske equation for the minimum number of plates Nb_{min} for two components, Nb_{min} has to be equal to C_B. Since the summation of all components must equal one ($\sum_{i=1}^{n} x_B^{(i)} = 1$), an equation can be found which relates the new bottom product composition to the current still composition in terms of relative volatilities and the minimum number of plates (Equation 6.8).

$$x_B^{(k)} = \frac{1}{\sum_{i=1}^{n} \left(\frac{\alpha_i}{\alpha_k}\right)^{-C_B} \frac{x_s^{(i)}}{x_s^{(k)}}} \qquad (6.8)$$

For a binary mixture, the separation is between only two components; in the case of a multicomponent mixture, however, the separation can be expressed in terms of a binary mixture of two key components, LK, the lightest component appearing in the bottom, and HK, the heavy key component defined as the heaviest component in the top. It should be noted that for most of the cases, the heavy key component in the stripper is the least volatile component in the mixture, as the still contains all the feed. Unlike continuous distillation, the key components keep changing for batch distillation as time progresses. For example, the light key component will become lighter than light key, and the next component in the relative volatility hierarchy will become the light key component.

To get a second boundary of the operating condition, the minimum reboil ratio Rb_{min}, Underwood's equations for continuous distillation columns are applied to batch distillation columns. At minimum reboil conditions, an infinite number of equilibrium stages is required to achieve the desired separation. The Underwood equations:

$$\sum_{i=1}^{n} \frac{\alpha_i x_F^{(i)}}{\alpha_i - \phi} = 1 - q \qquad (6.9)$$

$$-Rb_{min} = \sum_{i=1}^{n} \frac{\alpha_i x_B^{(i)}}{\alpha_i - \phi} \qquad (6.10)$$

The q in the above equations represents the feed condition and is defined as the ratio of heat required to vaporize 1 mol of the feed to the molar latent heat of the feed. The ϕ is the root of the Underwood equation which lies between α_{LK} and α_{HK}. The shortcut method assumes that batch distillation can be considered as continuous distillation with changing feed. In other words, the top product of one time step forms the feed for the next time step. This is equivalent to having the top plate as the feed plate and the feed at its boiling point, which means q is unity. Also, the feed composition in the Underwood equations can be replaced by still composition. Therefore, the Underwood equations for the batch stripper are given by:

The Underwood equations:

$$\sum_{i=1}^{n} \frac{\alpha_i x_s^{(i)}}{\alpha_i - \phi} = 0 \qquad (6.11)$$

$$-Rb_{min} = \sum_{i=1}^{n} \frac{\alpha_i x_B^{(i)}}{\alpha_i - \phi} \qquad (6.12)$$

The Fenske and Underwood equations above provide the limiting boundary conditions for a batch stripper in terms of the still and bottom composition. These limiting conditions should be related to the design variables of the column such as the reboil ratio Rb and the number of plates N_B to complete the analysis. In a batch rectifier, Gilliland's correlation furnishes this information for relating the design variables N and R with R_{min} and N_{min}. For continuous distillation, from which the shortcut procedure for the rectifier is derived, FUG equations supply the complete design equations including the stripping section of the column. This is because in continuous distillation the rectifying and stripping section, and hence reflux and reboil ratios are connected to each other by steady state material balance equations. However, for the batch stripper or for the middle vessel column, the steady state balance equation does not exist. Therefore, a correlation similar to Gilliland's correlation needs to be obtained for the stripper. Lotter and Diwekar(1997) obtained this correlation by conducting a large number of systematic experiments with various systems. We call this correlation (Equation 6.13) the modified Gilliland correlation. It can be seen that unlike the original Gilliland correlation, this correlation includes the effect of relative volatility.

Modified Gilliland Correlation:

$$Y = 0.2478 - 0.0965 \ln(3.784X) \qquad (6.13)$$

where

$$X = \frac{Rb - Rb_{min}}{Rb} \ln(\alpha_{LK}/\alpha_{HK}); \qquad Y = \frac{N_B - Nb_{min}}{N_B + 1}$$

The above correlation completes the shortcut model for the batch stripper. The following procedure illustrates the shortcut method for a single time step in detail for the constant reboil mode of operation. Note that choosing the proper light key and heavy key components is crucial for the success of this method similar to the batch rectifier shortcut method.

Constant Reboil Mode: At any time step the still composition, $x_s^{(i)}$, $i = 1, 2, \ldots, n$ can be found from the previous time steps using Equation 6.5. Then the following procedure is used to obtain the bottom compositions.

1. Assume the initial value of C_B ($0 < C_B \le Nb$).

2. Calculate the bottom product composition of the key component k using the HG equation and summation of all compositions.

$$x_B^{(k)} = \frac{1}{\sum_{i=1}^{n} \left(\frac{\alpha_i}{\alpha_k}\right)^{-C_B} \frac{x_s^{(i)}}{x_s^{(k)}}}$$

3. Substitute the value in the Hengestebeck–Geddes equation and find the bottom composition, $x_B^{(i)}$, $i = 1, 2, \ldots, n$.

$$x_B^{(i)} = \left(\frac{\alpha_i}{\alpha_k}\right)^{-C_B} \frac{x_B^{(k)}}{x_s^{(k)}} x_s^{(i)}, i = 1, 2, \ldots, n (i \ne k)$$

4. Solve the Underwood equations for ϕ and obtain the value of Underwood's minimum reboil ratio, Rb_{minu}.

$$\sum_{i=1}^{n} \frac{\alpha_i x_s^{(i)}}{\alpha_i - \phi} = 0$$

$$-Rb_{minu} = \sum_{i=1}^{n} \frac{\alpha_i x_B^{(i)}}{\alpha_i - \phi}$$

5. Calculate Y using Nb_{min} equal to C_B and solve the modified Gilliland correlation for the stripper for X to obtain Gilliland's minimum reboil ratio, Rb_{ming}.

$$X = \frac{Rb - Rb_{ming}}{Rb} ln(\alpha_{LK}/\alpha_{HK}); \quad Y = \frac{N_B - C_B}{N_B + 1}$$

$$Y = 0.2478 - 0.0965 ln(3.784X)$$

6. For the correct value of C_B, Rb_{minu} should be equal to Rb_{ming}. Therefore, the value of the quantity Gb_c should be zero (within a tolerance).

$$Gb_c = \frac{Rb_{minu} - Rb_{ming}}{Rb}$$

7. Calculate Gb_c and find whether it is zero within a tolerance. If Gb_c is approximately zero, the solution is converged for this time step; otherwise use the Newton-Raphson method to calculate the new C_B, and repeat all the steps from step 2.

The shortcut method for a batch stripper was compared with the rigorous model of a batch stripper for various systems by Lotter and Diwekar (1997). It has been found that the method is reasonably accurate and very efficient. Further, it can provide feasibility information for design, optimization, and optimal control problems.

6.1.2   Rigorous Model

The model is based on the assumptions of negligible vapor holdup, adiabatic operation, theoretical plates, and constant molar holdup.

The balance at the top of the column (Figure 6.4), i.e., for the stripper still, is given below.
Overall material balance:

$$\frac{dS}{dt} = V_1 - L_0 \tag{6.14}$$

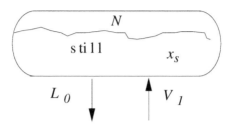

FIGURE 6.4
Top of the Column Balances

Mass balance of component i:

$$\frac{dx_s^{(i)}}{dt} = V_1/S(y_1^{(i)} - x_s^{(i)}) \tag{6.15}$$

Energy balance equations:

$$\frac{dSI_0}{dt} = V_1J_1 - L_0I_0 - Q_s \tag{6.16}$$

For an arbitrary plate j, the total mass, the component, and the finite-difference energy balances are given below.

The component balance for component i around plate j (Figure 6.5) is given by

$$\frac{dx_j^{(i)}}{dt} = \frac{1}{H_j}[V_{j+1}(y_{j+1}^{(i)} - x_j^{(i)}) + L_{j-1}(x_{j-1}^{(i)} - x_j^{(i)}) - V_j(y_j^{(i)} - x_j^{(i)})] \tag{6.17}$$

and from energy balance considerations:

$$V_{j+1} = \frac{1}{J_{j+1} - I_j}[V_j(J_j - I_j) + L_{j-1}(I_j - I_{j-1}) + H_j\delta I_j] \tag{6.18}$$

$$L_j = V_{j+1} + L_{j-1} - V_j - \delta H_j \tag{6.19}$$

where $j = 0, 1, \ldots, N$ and $i = 1, 2, \ldots, n$, and

- H_j is the molar holdup on plate j [mole]

- L_j is the liquid stream leaving plate j [mole/hr]

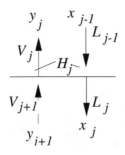

FIGURE 6.5
Plate Balances

- n is the number of components

- N is the number of plates in the column

- Q_s still heat duty [J/mole]

- V_j is the vapor stream leaving plate j [mole/hr]

- $x_j^{(i)}$ is the liquid composition of the component i on plate j [mole fraction]

- $y_j^{(i)}$ is the vapor composition of the component i which is in equilibrium with the liquid on plate j [mole fraction]

- I_j is the enthalpy of the liquid stream leaving plate j [J/mole]

- J_j is the enthalpy of the vapor stream leaving plate j [J/kmole]

The balance at the bottom of the column (Figure 6.6), i.e., from the reboiler, and express the reboil ratio as Rb,

$$Rb = \frac{V_B}{dB/dt} \tag{6.20}$$

Overall material balance:

$$\frac{dH_B}{dt} = L_N - V_B - \frac{dB}{dt} \tag{6.21}$$

Mass balance of component i:

$$\frac{dH_B x_B^{(i)}}{dt} = L_N x_N^{(i)} - V_B y_B^{(i)} - x_B^{(i)} \frac{dB}{dt} \tag{6.22}$$

Energy balance equations:

$$\frac{dH_B I_B}{dt} = L_N I_N - V_B J_B - I_B \frac{dB}{dt} - Q_B \tag{6.23}$$

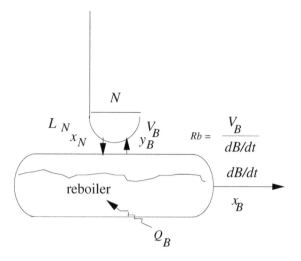

FIGURE 6.6
Reboiler Balances

Applying the same finite difference assumptions for the change in holdup and change in enthalpy as given in Chapter 3, Equations 6.21 and 6.22 transform into the following equations.
The liquid flowrate at the bottom of the column:

$$L_N = V_B \frac{(Rb+1)}{Rb} + \delta_t H_B \qquad (6.24)$$

The differential material balance for component i:

$$\frac{dx_B^{(i)}}{dt} = \frac{1}{H_B}[L_N(x_N^{(i)} - x_B^{(i)}) - V_B(y_B^{(i)} - x_B^{(i)})] \qquad (6.25)$$

Substituting Equation 6.24 in Equation 6.23 results in the following expression for the reboiler duty:

$$Q_B = L_N(I_N - I_B) - V_B(J_B - I_B) - H_B \delta_t I_B \qquad (6.26)$$

where

- H_B is the amount of material in the reboiler [mole]

- dB/dt is the bottom rate [mole/hr]

- I_B is the enthalpy of the liquid in the reboiler [J/mole]

- V_B is the vapor stream leaving the reboiler [mole/hr]

- $x_B^{(i)}$ is the liquid composition of component i in the reboiler [mole fraction]

- $y_B^{(i)}$ is the vapor composition of component i which is in equilibrium in the reboiler [mole fraction]

Given the values of the reboil ratio Rb, the number of plates, and the startup conditions described later in this section, the differential Equations 6.15, 6.17, and 6.25 can be integrated numerically to find the composition of the components in the still, reboiler, and on each plate. The flowrates are calculated starting from Equations 6.18 and 6.19. The heat duties are computed using Equations 6.16 and 6.26. The thermodynamics provide the equilibrium and the enthalpy relations for each plate and for the condenser and reboiler. Table 6.1 lists all the equations involved in the dynamic analysis of the batch stripper and the assumptions behind these equations. Initial conditions for these equations are given below.

Startup Conditions for the Batch Stripper:

Similar to the rectifier, this column can also be started with an initial total reflux operation consisting of the three steps outlined below.

1. Preheat the still charge to its bubble point.
2. Fill the column, still, and the reboiler holdups with the feed, so that the initial conditions for the total reboil operation may be written as follows:

$$x_s^{(i)} = x_j^{(i)} = x_F^{(i)}, i = 1, 2, \ldots, n; j = 1, 2, \ldots, N. \qquad (6.27)$$

3. Run without bottom withdrawal and take the unit to a steady state.

Again, there are two ways to simulate the startup conditions. One way is to use the steady state balances shown in Table 6.2 at the end of total reboil conditions and the other way is to simulate the transient equations shown in Table 6.3 using the initial conditions given above.

6.1.3 Semirigorous Model

The low-holdup semirigorous model for the batch stripper is given in Table 6.4.

6.2 Middle Vessel Column

The middle vessel column was proposed by Devidyan et al. (Devidyan et al., 1994). This column consists of a rectifier and an inverted column connected by a still vessel (Figure 6.2). Hasebe (Hasebe et al., 1995) describes this connection as a heat integration for the rectifier and stripper. This configuration

TABLE 6.1

Complete Column Dynamics for the Batch Stripper

Assumptions:
• Negligible vapor holdup   • Adiabatic operation
• Theoretical plates   • Constant molal holdup
• Finite difference approximations for enthalphy changes

Composition Calculations

Still Dynamics

$$\frac{dx_s^{(i)}}{dt} = V_1/S(y_1^{(i)} - x_s^{(i)}),$$
$$i = 1, 2, \ldots, n$$

Plate Dynamics

$$\frac{dx_j^{(i)}}{dt} = \frac{1}{H_j}[V_{j+1}y_{j+1}^{(i)} + L_{j-1}x_{j-1}^{(i)} - V_jy_j^{(i)} - L_jx_j^{(i)}],$$
$$i = 1, 2, \ldots, n; j = 1, 2, \ldots, N$$

Reboiler Dynamics

$$\frac{dx_B^{(i)}}{dt} = \frac{1}{B}[L_N(x_N^{(i)} - x_B^{(i)}) - V_B(y_B^{(i)} - x_B^{(i)})]$$

Flowrate Calculations

At the Top of the Column

$$\frac{dS}{dt} = V_1 - L_0$$

On the Plates

$$L_j = V_{j+1} + L_{j-1} - V_j; j = 1, 2, \ldots, N$$
$$V_{j+1} = \frac{1}{J_{j+1} - I_j}[V_j(J_j - I_j) + L_{j-1}(I_j - I_{j-1}) + H_j\delta I_j]$$

At the Bottom of the Column

$$L_N = V_B(Rb + 1)/Rb$$
$$\frac{dB}{dt} = \frac{V_B}{Rb}$$

Heat Duty Calculations

Still Duty

$$Q_s = V_1/S(J_1 - I_0) - \delta I_0$$

Reboiler Duty

$$Q_B = V_B((Rb + 1)/Rb(I_N - J_B) - dB/dt\, I_B - H_B\delta_t I_B$$

Thermodynamics Models

Equilibrium Relations

$$y_j^{(i)} = f((x_j^{(k)}, k = 1, \ldots, n), TE_j, P_j)$$

Enthalpy Calculations

$$I_j = f((x_j^{(k)}, j = 1, \ldots, n), TE_j, P_j)$$
$$J_j = f((y_j^{(k)}, j = 1, \ldots, n), TE_j, P_j)$$

TABLE 6.2
Steady State at the End of the Total Reboil Operation for the Batch Stripper

Composition Calculations
Still Balance
$0 = [x_0^{(i)} - y_1^{(i)}]$
Condenser Balance
Plate-to-Plate Calculations
$0 = [V_{j+1}y_{j+1}^{(i)} + L_{j-1}x_{j-1}^{(i)} - V_j y_j^{(i)} - L_j x_j^{(i)}],$ $i = 1, 2, \ldots, n; j = 1, 2, \ldots, N$
Reboiler Balance
$0 = [x_N^{(i)} - y_B^{(i)}]$
Overall Component Balance
$Fx_F^{(i)} = [(S)x_s^{(i)} + \sum_{j=1}^N H_j x_j^{(i)}]$
Flowrate Calculations
Reboiler
$L_N = V_B$
On the Plates
$L_j = V_{j+1} + L_{j-1} - V_j; j = 1, 2, \ldots, N$ $V_{j+1} = \frac{1}{J_{j+1} - I_j}[V_j(J_j - I_j) + L_{j-1}(I_{j-1} - I_j)]$ $j = 1, 2, \ldots, N+1$
Thermodynamics Correlations
Equilibrium Relations
$y^{(i)} = f(x_j^{(k)}, k = 1, \ldots, n, TE_j, P_j)$
Enthalpy Calculations
$I_j = f((x_j^{(k)}, j = 1, \ldots, n), TE_j, P_j)$ $J_j = f((y_j^{(k)}, j = 1, \ldots, n), TE_j, P_j)$

TABLE 6.3

The Startup Operation Dynamic Model for the Batch Stripper

Composition Calculations
Still Dynamics
$\frac{dx_s^{(i)}}{dt} = V_1/S(y_1^{(i)} - x_s^{(i)}),$
$i = 1, 2, \ldots, n$
Plate Dynamics
$\frac{dx_j^{(i)}}{dt} = \frac{1}{H_j}[V_{j+1}y_{j+1}^{(i)} + L_{j-1}x_{j-1}^{(i)} - V_jy_j^{(i)} - L_jx_j^{(i)}],$
$i = 1, 2, \ldots, n; j = 1, 2, \ldots, N$
Reboiler Dynamics
$\frac{dx_B^{(i)}}{dt} = \frac{1}{B}[L_N(x_N^{(i)} - x_B^{(i)}) - V_B(y_B^{(i)} - x_B^{(i)})]$
Flowrate Calculations
On the Plates
$L_j = V_{j+1} + L_{j-1} - V_j; j = 1, 2, \ldots, N$
$V_{j+1} = \frac{1}{J_{j+1} - I_j}[V_j(J_j - I_j) + L_{j-1}(I_j - I_{j-1}) + H_j\delta I_j]$
At the Bottom of the Column
$L_N = V_B$
Heat Duty Calculations
Sill Duty
$Q_s = V_1/S(J_1 - I_0) - \delta I_0$
Reboiler Duty
$Q_B = V_B(I_N - J_B) - H_B\delta_t I_B$
Thermodynamics Models
Equilibrium Relations
$y_j^{(i)} = f((x_j^{(k)}, k = 1, \ldots, n), TE_j, P_j)$
Enthalpy Calculations
$I_j = f((x_j^{(k)}, j = 1, \ldots, n), TE_j, P_j)$
$J_j = f((y_j^{(k)}, j = 1, \ldots, n), TE_j, P_j)$

TABLE 6.4
The Semirigorous Model for the Batch Stripper

Composition Calculations
Still Dynamics
$\frac{dx_s^{(i)}}{dt} = V_1/S(y_1^{(i)} - x_s^{(i)}),$ $i = 1, 2, \ldots, n$ $\frac{dS}{dt} = V_1 - L_0 ,$ *Plate Dynamics* $0 = [V_{j+1}y_{j+1}^{(i)} + L_{j-1}x_{j-1}^{(i)} - V_j y_j^{(i)} - L_j x_j^{(i)}],$ $i = 1, 2, \ldots, n; j = 1, 2, \ldots, N$ *Reboiler Dynamics* $0 = [L_N(x_N^{(i)} - x_B^{(i)}) - V_B(y_B^{(i)} - x_B^{(i)})]$
Flowrate Calculations
On the Plates
$L_j = V_{j+1} + L_{j-1} - V_j; j = 1, 2, \ldots, N$ $V_{j+1} = \frac{1}{J_{j+1} - I_j}[V_j(J_j - I_j) + L_{j-1}(I_j - I_{j-1})]$ *At the Bottom of the Column* $L_N = V_B(Rb+1)/Rb$ $\frac{dB}{dt} = V_B/Rb$
Heat Duty Calculations
Sill Duty
$Q_s = V_1/S(J_1 - I_0) - \delta I_0$ *Reboiler Duty* $Q_B = V_B((Rb+1)(I_N - J_B) - dB/dt\, I_B$
Thermodynamics Models
Equilibrium Relations
$y_j^{(i)} = f((x_j^{(k)}, k = 1, \ldots, n), TE_j, P_j)$ *Enthalpy Calculations* $I_j = f((x_j^{(k)}, j = 1, \ldots, n), TE_j, P_j)$ $J_j = f((y_j^{(k)}, j = 1, \ldots, n), TE_j, P_j)$

enables the simultaneous separation of light and heavy impurities and therefore offers more flexibility.

The middle vessel column makes it possible to copy the behavior of a column operated at variable reflux without having to change the reflux or reboil ratio. The goal of the variable reflux operation is to maintain a specified product composition by changing the reflux ratio. This can be achieved in a middle vessel column by keeping the composition of the middle vessel constant. This means that the vapor boilup rates for the top and bottom parts of the column and the reflux and reboil ratio have to be designed in a way to make up the difference in the mass balance equation. Since this column consists of a rectifier and an inverted column with both connected by the reservoir in the middle, its behavior becomes less predictable than a conventional column.

Figure 6.7 represents a schematic of a batch distillation column with a middle vessel. The following subsection presents various models for the middle vessel column (MVC).

6.2.1   Shortcut Method

The shortcut model for the middle vessel column (Lotter and Diwekar, 1997) is derived from the rectifier and the stripper shortcut model and is described below.

Since the mass balance around the middle vessel constitutes the main difference between a middle vessel column, an inverted column, and a rectifier, it is responsible for the change in column behavior.

$$\frac{dH_m}{dt} = \frac{V_B}{Rb} - \frac{V_T}{R+1}, \quad S_0 = F \qquad (6.28)$$

$$\frac{dB}{dt} = \frac{V_B}{Rb}, \quad B_0 = 0.0 \qquad (6.29)$$

$$x_B^{(k)}\frac{V_B}{Rb} + x_D^{(k)}\frac{V_T}{R+1} = -\frac{d(x_m^{(k)}H_m)}{dt}, \quad x_m^{(k)} = x_F(k) \quad (6.30)$$

$$\frac{dx_m^{(k)}}{dt} = \frac{V_B}{(Rb)S}(x_m^{(k)} - x_B^{(k)}) + \frac{V_T}{(R+1)H_m}(x_m^{(k)} - x_D^{(k)}), \quad (6.31)$$

$$x_{m0}^{(k)} = x_F^{(k)}$$

where V_T is the vapor boilup rate for the top part of the column and $x_D^{(k)}$ is the distillate composition of the key component. The combination of the overall and component mass balance leads to an equation that updates the middle vessel composition in terms of the distillate and the bottom product composition.

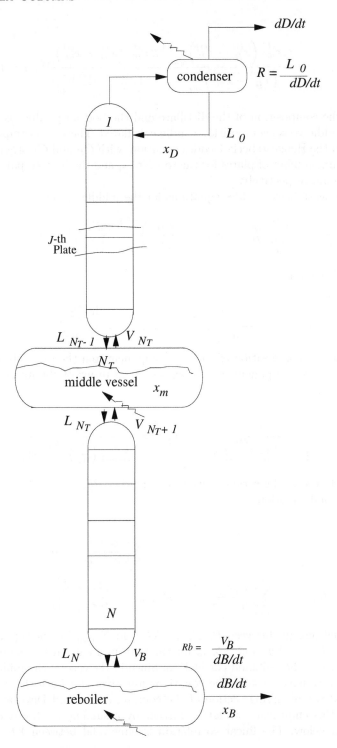

FIGURE 6.7
Column with a Middle Vessel

$$x^{(i)}_{m_{new}} = x^{(i)}_{m_{old}} + \frac{\triangle x^{(k)}_m \left(x^{(i)}_B - x^{(i)}_m\right)_{old}}{\left(x^{(k)}_B - x^{(k)}_m\right)_{old}} + \frac{\triangle x^{(k)}_m \left(x^{(i)}_D - x^{(i)}_m\right)_{old}}{\left(x^{(k)}_D - x^{(k)}_m\right)_{old}}, \quad i = 1, 2, \ldots, i \neq k$$

(6.32)

Once the composition of the distillate and the bottom product is known, the new middle vessel composition can be calculated. These two compositions result from the Hengestebeck–Geddes equation, with C_B and C_T standing for the minimum number of plates for the top Nt_{min} and the bottom part Nb_{min} of the column, respectively.

The Hengestebeck–Geddes equations for the middle vessel

$$x^{(i)}_D = \left(\frac{\alpha_i}{\alpha_k}\right)^{C_T} \frac{x^{(i)}_m}{x^{(k)}_m} x^{(k)}_D, \quad x^{(i)}_B = \left(\frac{\alpha_i}{\alpha_k}\right)^{-C_B} \frac{x^{(i)}_m}{x^{(k)}_m} x^{(k)}_B, \quad i = 1, 2, \ldots, n$$

(6.33)

Fenske's equation:

$$Nt_{min} = \frac{\ln\left[\frac{x^{(i)}_D}{x^{(k)}_D} \frac{x^{(i)}_m}{x^{(k)}_m}\right]}{\ln\left[\frac{\alpha_i}{\alpha_k}\right]}, \quad Nb_{min} = \frac{\ln\left[\frac{x^{(i)}_m}{x^{(k)}_m} \frac{x^{(i)}_B}{x^{(k)}_B}\right]}{\ln\left[\frac{\alpha_i}{\alpha_k}\right]}$$

(6.34)

The bottom composition of the key component and the distillate composition of the key component can be expressed in terms of the middle vessel composition.

$$x^{(k)}_D = \frac{1}{\sum_{i=1}^{n} \left(\frac{\alpha_i}{\alpha_k}\right)^{C_T} \frac{x^{(i)}_m}{x^{(k)}_m}}, \quad x^{(k)}_B = \frac{1}{\sum_{i=1}^{n} \left(\frac{\alpha_i}{\alpha_k}\right)^{-C_B} \frac{x^{(i)}_m}{x^{(k)}_m}}$$

(6.35)

Similarly the Underwood equations can predict Rb_{min} and R_{min}. Underwood equations:

$$\sum_{i=1}^{n} \frac{\alpha_i x^{(i)}_s}{\alpha_i - \phi} = 0$$

(6.36)

$$R_{min} = \sum_{i=1}^{n} \frac{\alpha_i x^{(i)}_D}{\alpha_i - \phi} \quad - Rb_{min} = \sum_{i=1}^{n} \frac{\alpha_i x^{(i)}_B}{\alpha_i - \phi}$$

(6.37)

The relationships between R, R_{min}, Nt, and Nt_{min} for the top, and between Rb, Rb_{min}, Nb, and Nb_{min} for the bottom are given by an empirical relation such as the Gilliland correlation. Since in the case of the middle vessel column, we have to solve the equations iteratively for both the stripping section and the rectifying section of the column, Lotter and Diwekar (1997) simplified this empirical correlation further by assuming the linear correlations given below. The linear correlation is only valid between $Yt \leq 0.6$

for the rectifier and $Yb \leq 0.55$. For wider applicability one can always use Gilliland's correlation for the rectifying section and Equation 6.13 for the stripping section.

$$Yt = 0.5515 - 0.5948Xt \qquad (6.38)$$

$$Yb = 0.6187 - 0.5655Xb \qquad (6.39)$$

where

$$Xt = \frac{R - R_{min}}{R + 1}; \quad Yt = \frac{N_T - Nt_{min}}{Nt + 1}$$

$$Xb = \frac{Rb - Rb_{min}}{Rb} \ln \alpha_{LK}/\alpha_{HK}; \quad Yb = \frac{N_B - Nb_{min}}{Nb + 1}$$

$$N_B = N - N_T$$

Thus Equations 6.28 to 6.39 form the shortcut model for the middle vessel column. At each time step the model needs to solve three differential equations (Equations 6.28, 6.29, and 6.32) and the algebraic equations consisting of Equation 6.32, HG equations, and FUG equations iteratively. The following subsection describes this procedure for the constant reflux/constant reboil mode of operation.

Constant Reflux/Constant Reboil Mode: At any time step the middle vessel composition, $x_m^{(i)}$, $i = 1, 2, \ldots, n$, can be found from the previous time steps using Equation 6.32. Then the following procedure is used to obtain the bottom and top compositions.

1. Assume the initial value of C_T and C_B. Remember that C_T and C_B should be greater than zero but less than the number of plates in the respective sections.

2. Calculate the distillate and bottom product composition of key component k by using the corresponding Hengestebeck–Geddes equation and summation of all compositions.

$$x_D^{(k)} = \frac{1}{\sum_{i=1}^{n} \left(\frac{\alpha_i}{\alpha_k}\right)^{C_T} \frac{x_m^{(i)}}{x_m^{(k)}}}, \quad x_B^{(k)} = \frac{1}{\sum_{i=1}^{n} \left(\frac{\alpha_i}{\alpha_k}\right)^{-C_B} \frac{x_m^{(i)}}{x_m^{(k)}}}$$

3. Find all other distillate and bottom compositions by using the HG equations.

$$x_D^{(i)} = \left(\frac{\alpha_i}{\alpha_k}\right)^{C_T} \frac{x_D^{(k)}}{x_m^{(k)}} x_m^{(i)}, \quad x_B^{(i)} = \left(\frac{\alpha_i}{\alpha_k}\right)^{-C_B} \frac{x_B^{(k)}}{x_m^{(k)}} x_m^{(i)}, i = 1, 2, \ldots, n(i \neq k)$$

4. Solve the Underwood equations for ϕ and obtain the value of Underwood's minimum reflux ratio, R_{minu}, and minimum reboil ratio, Rb_{minu}.

$$\sum_{i=1}^{n} \frac{\alpha_i x_s^{(i)}}{\alpha_i - \phi} = 0$$

$$R_{minu} = \sum_{i=1}^{n} \frac{\alpha_i x_D^{(i)}}{\alpha_i - \phi}, \qquad -Rb_{minu} = \sum_{i=1}^{n} \frac{\alpha_i x_B^{(i)}}{\alpha_i - \phi}$$

5. Calculate Yt and Yb using Nt_{min} (equivalent to C_T) and Nb_{min} (equivalent to C_B) and solve the modified Gilliland correlations for the Xt and Xb, respectively.

$$Xt = \frac{R - R_{ming}}{R + 1}; \qquad Yt = \frac{Nt - C_T}{N_T + 1}$$

$$Yt = 0.5515 - 0.5948 Xt$$

$$Xb = \frac{Rb - Rb_{ming}}{Rb} \ln \alpha_{LK}/\alpha_{HK}; \qquad Yb = \frac{N_B - C_B}{N_B + 1}$$

$$Yb = 0.6187 - 0.5655 Xb$$

6. For the correct values of C_T, R_{minu} should be equal to R_{ming}, hence, Gt_c should be equal to zero. Similarly, Rb_{minu} should be equal to Rb_{ming}. Therefore, the value of the quantity Gb_c should be zero.

$$Gt_c = \frac{R_{minu} - R_{ming}}{R + 1}$$

$$Gb_c = \frac{Rb_{minu} - Rb_{ming}}{Rb}$$

7. For the top column iterate on C_T until Gt_c is zero within a tolerance and for the bottom portion of the column iterate on C_B for Gb_c to be negligible.

6.2.2 Rigorous Model

The model is based on the assumptions of negligible vapor holdup, adiabatic operation, theoretical plates, and constant molar holdup. For an arbitrary plate j, the total mass, the component, and the finite-difference energy balances as used in Chapter 3 for the rectifier yield the following equations for the plate and condenser balances (the detailed derivations of these equations are given in Chapter 3).

Component balance for component i around plate j (Figure 6.8) is given by

$$\frac{dx_j^{(i)}}{dt} = \frac{1}{H_j}[V_{j+1}(y_{j+1}^{(i)} - x_j^{(i)}) + L_{j-1}(x_{j-1}^{(i)} - x_j^{(i)}) - V_j(y_j^{(i)} - x_j^{(i)})] \quad (6.40)$$

and from energy balance considerations:

$$V_{j+1} = \frac{1}{J_{j+1} - I_j}[V_j(J_j - I_j) + L_{j-1}(I_j - I_{j-1}) + H_j\delta I_j] \quad (6.41)$$

$$L_j = V_{j+1} + L_{j-1} - V_j - \delta H_j \quad (6.42)$$

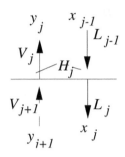

FIGURE 6.8
Plate Balances

where $j = 0, 1, \ldots, N$; $j \neq N_T$ and $i = 1, 2, \ldots, n$, $N = N_T + N_B$, and

- H_j is the molar holdup on plate j [mole]

- L_j is the liquid stream leaving plate j [mole/hr]

- n is the number of components

- $N_T - 1$ is the number of plates in the top section of the column

- N_B is the number of plates in the bottom section of the column

- N_T is the plate corresponding to the middle vessel

- V_j is the vapor stream leaving plate j [mole/hr]

- $x_j^{(i)}$ is the liquid composition of component i on plate j [mole fraction]

- $y_j^{(i)}$ is the vapor composition of component i which is in equilibrium with the liquid on plate j [mole fraction]

- I_j is the enthalpy of the liquid stream leaving plate j [J/mole]

- J_j is the enthalpy of the vapor stream leaving plate j [J/kmole]

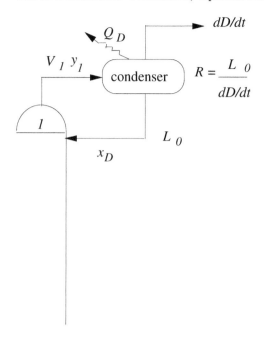

FIGURE 6.9
Condenser Balances

Figure 6.9 shows the balances at the top of the column around the condenser. The vapor and liquid flowrates at the top of the column are

$$L_0 = R\frac{dD}{dt} \tag{6.43}$$

$$V_1 = \frac{dD}{dt}(R+1) + \delta_t H_D \tag{6.44}$$

The differential balance for the distillate composition of component i is

$$\frac{dx_D^{(i)}}{dt} = \frac{V_1}{H_D}(y_1^{(i)} - x_D^{(i)}) \tag{6.45}$$

The condenser heat duty Q_D is

$$Q_D = V_1(J_1 - I_D) - H_D\delta_t I_D \tag{6.46}$$

where

- dD/dt is the distillate rate [mole/hr]

- $H_D = H_0$ is the molar condenser holdup [moles]

- I_D is the enthalpy of the liquid in the condenser [J/mole]

- L_0 is the liquid reflux at the top of the column [mole/hr]

- R is the reflux ratio

- $x_D^{(i)} = x_0^{(i)}$ is the composition of component i in the distillate [mole fraction]

In this column, the top section and the middle vessel are separate from the bottom section. Therefore, the balances, especially for calculation of flowrates, need to be considered starting from the top and ending at the middle vessel and starting from the bottom and again going up to the middle vessel.

Consider the balance at the bottom of the column (Figure 6.10), i.e., from the reboiler, and express the reboil ratio as Rb,

$$Rb = \frac{V_B}{dB/dt} \tag{6.47}$$

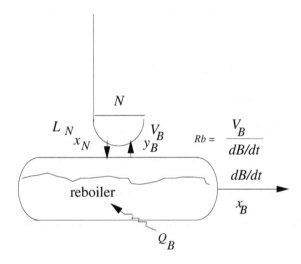

FIGURE 6.10
Reboiler Balances

Overall material balance:

$$\frac{dH_B}{dt} = L_N - V_B - \frac{dB}{dt} \tag{6.48}$$

Mass balance of component i:

$$\frac{dH_B x_B^{(i)}}{dt} = L_N x_N^{(i)} - V_B y_B^{(i)} - x_B^{(i)} \frac{dB}{dt} \tag{6.49}$$

Energy balance equations:

$$\frac{dH_B I_B}{dt} = L_N I_N - V_B J_B - I_B \frac{dB}{dt} - Q_B \tag{6.50}$$

Applying the same finite difference assumptions for the change in holdup and change in enthalpy as given in Chapter 3, Equations 6.48 and 6.49 transform into the following equations.

The liquid flowrate at the bottom of the column:

$$L_N = V_B \frac{(Rb+1)}{Rb} + \delta_t H_B \qquad (6.51)$$

The differential material balance for component i:

$$\frac{dx_B^{(i)}}{dt} = \frac{1}{H_B}[L_N(x_N^{(i)} - x_B^{(i)}) - V_B(y_B^{(i)} - x_B^{(i)})] \qquad (6.52)$$

Substituting Equation 6.51 in Equation 6.50 results in the following expression for the reboiler duty:

$$Q_B = L_N(I_N - I_B) - V_B(J_B - I_B) - H_B \delta_t I_B \qquad (6.53)$$

where

- H_B is the amount of material in the reboiler [mole]

- dB/dt is the bottom rate [mole/hr]

- I_B is the enthalpy of the liquid in the reboiler [J/mole]

- V_B is the vapor stream leaving the reboiler [mole/hr]

- $x_B^{(i)}$ is the liquid composition of component i in the reboiler [mole fraction]

- $y_B^{(i)}$ is the vapor composition of component i which is in equilibrium in the reboiler [mole fraction]

The plate balances for the bottom section of the column are the same as those given in Equations 6.40 to 6.42 except for the plate where the middle vessel is located. Figure 6.11 shows the middle vessel plate balance (the middle vessel plate is the (N_T)-th plate from the top in the column).

$$q' = \frac{V_{N_T}}{V_{N_T+1}} \qquad (6.54)$$

$$\frac{dH_m}{dt} = V_{N_T+1} + L_{N_T-1} - V_{N_T} - L_{N_T} \qquad (6.55)$$

Component balance for component i around the middle vessel:

$$\frac{dH_m x_m^{(i)}}{dt} = V_{N_T+1} y_{N_T+1}^{(i)} + L_{N_T-1} x_{N_T-1}^{(i)} - V_{N_T} y_m^{(i)} - L_{N_T} x_m^{(i)} \qquad (6.56)$$

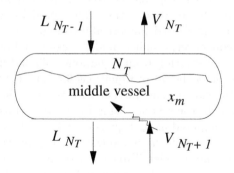

FIGURE 6.11
Middle Vessel Plate Balances

Energy balance around the middle vessel:

$$\frac{dH_m I_m}{dt} = V_{N_T+1} J_{N_T+1} + L_{N_T-1} I_{N_T-1} - V_{N_T} J_m - L_{N_T} I_m$$
$$- Q_m \qquad (6.57)$$

Substituting Equation 6.55 in Equation 6.56 results in the following equation:

$$\frac{dx_m^{(i)}}{dt} = \frac{1}{H_m}[V_{N_T+1}(y_{N_T+1}^{(i)} - x_m^{(i)}) + L_{N_T-1}(x_{N_T-1}^{(i)} - x_m^{(i)})$$
$$- V_{N_T}(y_m^{(i)} - x_m^{(i)})] \qquad (6.58)$$

The heat duty for the middle vessel:

$$Q_m = V_{N_T+1}(J_{N_T+1} - I_m) - L_{N_T-1}(I_{N_T-1} - I_m)$$
$$V_{N_T}(J_m - I_m) - H_m \delta_t I_m \qquad (6.59)$$

where

- H_m is the amount of material remaining in the middle vessel [mole]

- I_m is the enthalpy of the liquid in the middle vessel [J/mole]

- $x_m^{(i)}$ is the liquid composition of component i in the middle vessel [mole fraction]

- $y_m^{(i)}$ is the vapor composition of component i which is in equilibrium in the middle vessel [mole fraction]

- Q_m is the heat duty of the middle vessel [J/hr]

Given the values of q', dD/dt, the reboil ratio Rb, reflux ratio R, and the number of plates at the top and bottom of the column and the startup conditions described later in this section, differential Equations 6.40, 6.45, 6.52, and 6.58 can be integrated numerically to find the composition of the components in the condenser, reboiler, middle vessel, and on each plate. The flowrates are calculated starting from Equations 6.43 and 6.44 from the top in the recursive scheme of the set of Equations 6.41 and 6.42 until the middle vessel. Continuing the calculations further with the same set (6.41 and 6.42) from the middle vessel to the reboiler using Equation 6.54 as a starting point. The heat duties are computed using Equations 6.46, 6.53, and 6.59. The thermodynamics provide the equilibrium and the enthalpy relations for each plate and for the condenser and reboiler. Table 6.5 lists all the equations involved in the dynamic analysis of the batch column and the assumptions behind these equations.

Startup Conditions for the Middle Vessel Column:

Similar to the rectifier and the stripper, this column can also be started with an initial total reflux/total reboil operation consisting of the three steps outlined below.

1. Preheat the still charge to its bubble point.
2. Fill the column, still, and the reboiler and the condenser holdups with the feed, so that the initial conditions for the total reflux/total reboil operation may be written as follows:

$$x_B^{(i)} = x_j^{(i)} = x_F^{(i)}, i = 1, 2, \ldots, n; j = 1, 2, \ldots, N \qquad (6.60)$$

3. Run without distillate and bottom withdrawal and take the unit to a steady state.

Again, there are two ways to simulate the startup conditions. One way is to use the steady state balances shown in Table 6.6 at the end of total reflux and total reboil conditions and the other way is to simulate the transient equations shown in Table 6.7 using the initial conditions given above.

6.2.3   Semirigorous Model

The dynamic analysis of a rectifier presented in Chapter 3 showed that a low-holdup semirigorous model is necessary when the dynamic system becomes very stiff. The low-holdup semirigorous model for the middle vessel column is given in Table 6.8.

We have seen the middle vessel complete dynamic model and the low-holdup semirigorous model. This column is very flexible and hence very effective. Practically, one can extract very pure components in the top, bottom, and the middle vessel. As one adds more features to this column, the column offers more flexibility to obtain the products of interest. For example,

TABLE 6.5

Complete Column Dynamics for the Middle Vessel Column

Assumptions:
• Negligible vapor holdup   • Adiabatic operation   • Theoretical plates   • Constant molal holdup   • Finite difference approximations for enthalphy changes

Composition Calculations
Condenser and Accumulator Dynamics

$$\frac{dx_D^{(i)}}{dt} = \frac{V_1}{H_D}(y_1^{(i)} - x_D^{(i)}), i = 1, 2, \ldots, n$$

Plate Dynamics (including middle vessel)

$$\frac{dx_j^{(i)}}{dt} = \frac{1}{H_j}[V_{j+1}y_{j+1}^{(i)} + L_{j-1}x_{j-1}^{(i)} - V_j y_j^{(i)} - L_j x_j^{(i)}],$$
$$i = 1, 2, \ldots, n; j = 1, 2, \ldots, N$$

Reboiler Dynamics

$$\frac{dx_B^{(i)}}{dt} = \frac{1}{B}[L_N(x_N^{(i)} - x_B^{(i)}) - V_B(y_B^{(i)} - x_B^{(i)})]$$

Flowrate Calculations
At the Top of the Column

$$L_0 = R\frac{dD}{dt}$$
$$V_1 = (R + 1)\frac{dD}{dt}$$

On the Plates

$$L_j = V_{j+1} + L_{j-1} - V_j; j = 1, 2, \ldots, N$$
$$V_{j+1} = \frac{1}{J_{j+1} - I_j}[V_j(J_j - I_j) + L_{j-1}(I_j - I_{j-1}) + H_j\delta I_j]$$

At the Middle Vessel

$$q' = \frac{V_{NT}}{V_{NT+1}}$$
$$\frac{dH_m}{dt} = V_{N_T + 1} + L_{N_T-1} - V_{N_T} - L_{N_T}$$

At the Bottom of the Column

$$L_N = V_B(Rb + 1)/Rb$$
$$\frac{dB}{dt} = \frac{V_B}{Rb}$$

Heat Duty Calculations
Condenser Duty

$$Q_D = V_1(J_1 - I_D) - H_D\delta_t I_D$$

Reboiler Duty

$$Q_B = V_B((Rb + 1)(I_N - J_B) - dB/dt\, I_B - H_B\delta_t I_B$$

Middle Vessel Duty

$$Q_m = V_{N_T+1}(J_{N_T+1} - I_m) - L_{N_T-1}(I_{N_T-1} - I_m) + V_{N_T}(J_m - I_m)$$
$$- H_m\delta_t I_m$$

Thermodynamics Models
Equilibrium Relations

$$y_j^{(i)} = f((x_j^{(k)}, k = 1, \ldots, n), TE_j, P_j)$$

Enthalpy Calculations

$$I_j = f((x_j^{(k)}, j = 1, \ldots, n), TE_j, P_j)$$
$$J_j = f((y_j^{(k)}, j = 1, \ldots, n), TE_j, P_j)$$

TABLE 6.6
Steady State at the End of the Total Reflux/Reboil Operation for the Middle
Vessel Column

Composition Calculations
Condenser Balance
$0 = (y_1^{(i)} - x_D^{(i)}), i = 1, 2, \ldots, n$
Plate-to-Plate Calculations
$0 = [V_{j+1}y_{j+1}^{(i)} + L_{j-1}x_{j-1}^{(i)} - V_jy_j^{(i)} - L_jx_j^{(i)}],$ $i = 1, 2, \ldots, n; j = 1, 2, \ldots, N$
Reboiler Balance
$0 = [L_N(x_N^{(i)} - x_B^{(i)}) - V_B(y_B^{(i)} - x_B^{(i)})]$
Overall Component Balance
$Fx_F^{(i)} = [(H_D)x_D^{(i)} + \sum_{j=1}^{N} H_jx_j^{(i)} + Bx_B^{(i)}]$
Flowrate Calculations
At the Top of the Column
$V_1 = L_0$
On the Plates
$L_j = V_{j+1} + L_{j-1} - V_j; j = 1, 2, \ldots, N$
$q' = V_{NT}/V_{NT+1}$
$V_{j+1} = \frac{1}{J_{j+1} - I_j}[V_j(J_j - I_j) + L_{j-1}(I_{j-1} - I_j)]$ $j = 1, 2, \ldots, N+1$
Thermodynamics Correlations
Equilibrium Relations
$y^{(i)} = f(x_j^{(k)}, k = 1, \ldots, n, TE_j, P_j)$
Enthalpy Calculations
$I_j = f((x_j^{(k)}, j = 1, \ldots, n), TE_j, P_j)$
$J_j = f((y_j^{(k)}, j = 1, \ldots, n), TE_j, P_j)$

TABLE 6.7

The Startup Operation Dynamic Model for the Middle Vessel

Composition Calculations
Condenser and Accumulator Dynamics
$$\frac{dx_D^{(i)}}{dt} = \frac{V_1}{H_D}(y_1^{(i)} - x_D^{(i)}), i = 1, 2, \ldots, n$$
Plate Dynamics
$$\frac{dx_j^{(i)}}{dt} = \frac{1}{H_j}[V_{j+1}y_{j+1}^{(i)} + L_{j-1}x_{j-1}^{(i)} - V_jy_j^{(i)} - L_jx_j^{(i)}],$$ $$i = 1, 2, \ldots, n; j = 1, 2, \ldots, N$$
Reboiler Dynamics
$$\frac{dx_B^{(i)}}{dt} = \frac{1}{B}[L_N(x_N^{(i)} - x_B^{(i)}) - V_B(y_B^{(i)} - x_B^{(i)})]$$
Flowrate Calculations
At the Top of the Column
$$L_0 = V_1$$
On the Plates
$$L_j = V_{j+1} + L_{j-1} - V_j; j = 1, 2, \ldots, N$$
$$V_{j+1} = \frac{1}{J_{j+1} - I_j}[V_j(J_j - I_j) + L_{j-1}(I_j - I_{j-1}) + H_j\delta I_j]$$
At the Middle Vessel
$$q' = \frac{V_{NT}}{V_{NT+1}}$$
$$\frac{dH_m}{dt} = V_{N_T+1} + L_{N_T-1} - V_{N_T} - L_{N_T}$$
At the Bottom of the Column
$$L_N = V_B$$
$$\frac{dB}{dt} = 0.0$$
Heat Duty Calculations
Condenser Duty
$$Q_D = V_1(J_1 - I_D) - H_D\delta_t I_D$$
Reboiler Duty
$$Q_B = V_B(I_N - J_B) - H_B\delta_t I_B$$
Middle Vessel Duty
$$Q_m = V_{N_T+1}(J_{N_T+1} - I_m) - L_{N_T-1}(I_{N_T-1} - I_m) + V_{N_T}(J_m - I_m) - H_m\delta_t I_m$$
Thermodynamics Models
Equilibrium Relations
$$y_j^{(i)} = f((x_j^{(k)}, k = 1, \ldots, n), TE_j, P_j)$$
Enthalpy Calculations
$$I_j = f((x_j^{(k)}, j = 1, \ldots, n), TE_j, P_j)$$
$$J_j = f((y_j^{(k)}, j = 1, \ldots, n), TE_j, P_j)$$

TABLE 6.8

The Semirigorous Model

Composition Calculations
Condenser and Accumulator Dynamics $$0 = (y_1^{(i)} - x_D^{(i)}), i = 1, 2, \ldots, n$$ *Plate Dynamics* $$0 = [V_{j+1} y_{j+1}^{(i)} + L_{j-1} x_{j-1}^{(i)} - V_j y_j^{(i)} - L_j x_j^{(i)}],$$ $$i = 1, 2, \ldots, n; j = 1, 2, \ldots, N$$ *Reboiler Dynamics* $$0 = [L_N(x_N^{(i)} - x_B^{(i)}) - V_B(y_B^{(i)} - x_B^{(i)})]$$
Flowrate Calculations   *At the Top of the Column* $$L_0 = R\frac{dD}{dt}$$ $$V_1 = (R+1)\frac{dD}{dt}$$ *On the Plates* $$L_j = V_{j+1} + L_{j-1} - V_j; j = 1, 2, \ldots, N$$ $$V_{j+1} = \frac{1}{J_{j+1} - I_j}[V_j(J_j - I_j) + L_{j-1}(I_j - I_{j-1})]$$ *At the Middle Vessel* $$q' = \frac{V_{NT}}{V_{NT+1}}$$ $$\frac{dH_m}{dt} = V_{N_T} + 1 + L_{N_T - 1} - V_{N_T} - L_{N_T}$$ $$\frac{dx_m^{(i)}}{dt} = \frac{1}{H_m}[V_{N_T+1}(y_{N_T+1}^{(i)} - x_m^{(i)}) + L_{N_T-1}(x_{N_T-1}^{(i)} - x_m^{(i)}) - V_{N_T}(y_m^{(i)} - x_m^{(i)})]$$ *At the Bottom of the Column* $$L_N = V_B(Rb+1)$$ $$\frac{dB}{dt} = V_B/Rb$$
Heat Duty Calculations   *Condenser Duty* $$Q_D = V_1(J_1 - I_D)$$ *Reboiler Duty* $$Q_B = V_B((Rb+1)(I_N - J_B) - dB/dt\,I_B$$ *Middle Vessel Duty* $$Q_m = V_{N_T+1}(J_{N_T+1} - I_m) - L_{N_T-1}(I_{N_T-1} - I_m) + V_{N_T}(J_m - I_m) - H_m\delta_t I_m$$
Thermodynamics Models   *Equilibrium Relations* $$y_j^{(i)} = f((x_j^{(k)}, k = 1, \ldots, n), TE_j, P_j)$$ *Enthalpy Calculations* $$I_j = f((x_j^{(k)}, j = 1, \ldots, n), TE_j, P_j)$$ $$J_j = f((y_j^{(k)}, j = 1, \ldots, n), TE_j, P_j)$$

recently Safrit et al. (1995) investigated this column for extractive distillation and found that this column can recover all of the pure distillate product from an azeotropic feed where a rectifier alone would require a still-pot of infinite size. The following numerical example illustrates how interesting this column behavior can be with the proper values of design variables. The column obviously has more degrees of freedom and may require an optimizer in addition to the model to obtain the desired separation.

Example 6.1: A mixture containing 100 moles of a equimolar mixture of three components having relative volatilities of 1.44, 1.20, and 1.0 is to be distilled in a middle vessel batch distillation column. The column has five theoretical plates in the top section and five theoretical plates in the bottom section. The vapor boilup rate for the top of the column is 40 mole/hr and the vapor rate at the bottom of the column is 40 mols/hr. Assume equimolal overflow. The column is operating under a constant reflux mode with a reflux ratio equal to 5.0 and a reboil ratio equal to 5.0.

a) Simulate a 1 hr operation of the column and present the distillate composition profile.

b) Change the vapor rate at the top of the column to 10.5 and compare the results with (a).

Solution:

a) and b) Figure 6.12 shows the comparison of the profiles at different values of q'. It can be seen that by changing the value of q' one can obtain totally different distillate composition trends. Notice that the distillate composition of the more volatile component is increasing with time. As seen in the feasibility considerations (Chapter 5) for the rectifier, the more volatile component distillate composition is always highest in the beginning. However, for this operating condition the middle vessel shows exactly the opposite trend. This is indeed a very interesting behavior and can be attributed to the fact that the distillate composition is increasing because the less volatile component is pulled out in the bottom, making the composition of the more volatile component in the middle vessel more than the starting value.

6.3 Synthesis

The complexity in batch distillation design and operation is also reflected in the batch distillation synthesis problem. In continuous distillation, optimal column sequencing is the main focus of synthesis research. Several past reviews are available on this subject (Stephanopoulos, 1980; Nishida et al., 1981; Grossmann, 1990). Unlike continuous distillation synthesis, the area of batch distillation synthesis is complicated by transient nature. Decisions like cut selection, operating mode, configuration type, and column sequencing enter into the synthesis problem. For complex systems like azeotropic, extractive,

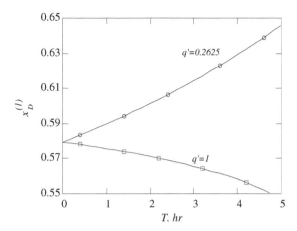

FIGURE 6.12
Transient Composition Profiles for Example 6.3b, Effect of q'

and reactive distillation, identifying the region boundaries and steering toward feasible and optimal regions add further complications to the problem.

The recent literature in batch distillation has been devoted to comparing emerging column configurations with the conventional one, thereby obtaining heuristics for optimal column configuration, optimal design, and optimal operating conditions (Chiotti and Iribarren, 1991; Meski and Morari, 1995; Hasebe et al., 1995; Sørensen and Skogestad, 1996; Kim and Diwekar, 2000). In these studies, parameters such as product purity, batch time, or total cost were evaluated to compare the performance of column configurations. Chiotti and Iribarren (1991) compared the rectifier with the stripper in terms of annual cost and product purity. They noted that the rectifier is better for the more volatile component products while it is more economical to obtain less volatile component products using the stripper. Meski and Morari (1995) compared three column configurations in terms of the batch time under fixed product purity and an infinite number of plates. It was observed that the middle vessel column always has the shortest batch time, and the rectifier is the next. Sørensen and Skogestad (1996) studied two competing column configurations, rectifier and stripper, in the context of minimum optimal operating time, and also provided the dynamic behavior of these columns. They concluded that the stripper is the preferred column configuration when a small amount of the more volatile component is in the feed, and that the rectifier is better when the feed has a high amount of the more volatile component. Although several studies support the same heuristics, there are also studies which present contradictions between suggested heuristics. For example, the batch time studies of Meski and Morari (1995) and Sørensen and Skogestad (1996) give a conflicting result with respect to feed composition. This is due to limited ranges of parameters and systems considered as well as the complexity and difficulty of the problem of column selection.

In order to elicit comprehensive heuristics, the analysis must cover a wider range of column configurations, operation policies, and design variables, and various performance indices need to be included. Kim and Diwekar (2000) extended the column configuration problem using four performance indices: product purity, yield, design feasibility and flexibility, and thermodynamic efficiency. It is generally observed that the rectifier is a promising column configuration for the more volatile component product, and that the stripper is better in the opposite case. Feasibility studies based on the minimum number of plates and the minimum reflux ratio showed whether such a high purity configuration is flexible for changing operating conditions. It was found that the rectifier and the stripper have distinctive feasibility regions in terms of the feed composition. Thermodynamic efficiency indicates how close a process or system is to its ultimate performance and also suggests whether the process or system can be improved or not. The rectifier can also be a promising column configuration in terms of thermodynamic efficiency, but in some conditions, higher efficiencies of the stripper or the middle vessel column are observed. Furthermore, for the middle vessel column, the thermodynamic efficiency is greatly affected by an added degree of freedom (q'). This systematic and parametric study concludes that the trade-offs between performance indices should be presented as a multi-objective framework.

References

Chiotti O. J. and O. A. Iribarren (1991), Simplified models for binary batch distillation, *Computers and Chemical Engineering*, **15**, 1.

Devidyan A. G., V. N. Kiva, G. A. Meski, and M. Morari (1994), Batch distillation in a column with a middle vessel, *Chemical Engineering Science*, **49**, 3033.

Diwekar U. M. (1991), An efficient design method for binary azeotropic batch distillation, *AIChE Journal*, **14**, 946.

Diwekar, U. M. and K. P. Madhavan (1991), Multicomponent batch distillation column design, *Ind. Eng Chem. Res.*, **30**, 713.

Grossmann I. E. (1990), MINLP optimization strategies and algorithms for process synthesis, *Proceedings of Foundations of Computer Aided Process Design 89*, Siirola et al. Eds., Elsevier, Amsterdam, The Netherlands.

Hasebe, S., T. Kurooka, and I. Hashimoto (1995), Comparison of the separation performances of a multi-effect batch distillation system and a continuous distillation system, *Proc. IFAC-Symposium DYCORD'95*, Denmark, June 1995.

Kim K. and U. Diwekar (2000), Comparing batch column configurations: a parametric study involving multiple objectives, *AIChE J.*, **46**, 2475.

Lotter S. and U. Diwekar (1997), Shortcut models and feasibility considerations for emerging batch distillation columns, *Ind. Eng. Chem. Res.*, **36 (3)**, 760.

Meski G. and M. Morari (1995), Design and operation of a batch distillation column with a middle vessel, *Comp. Chem. Eng.*, **19**, s597.

Nishida N., G. Stephanopoulos and A. Westerberg (1981), Review of process synthesis,*AIChE J.*, **27**, 321.

Safrit B., A. W. Westerberg, U. M. Diwekar, and O. M. Wahnschafft (1995), Insights into batch extractive distillation using a middle vessel column, *Industrial and Engineering Chemistry Research*, **34**, 3257.

Sørensen E. and S. Skogestad (1996), Comparison of regular and inverted batch distillation,*Chem. Eng. Sci.*, **51**, 4949.

Skogestad S., B. Wittgens, E. Sørensen, and R. Litto (1995), Multivessel batch distillation, *paper presented at the AIChE Annual Meeting*, Miami, FL.

Skogestad S., B. Wittgens, R. Litto, and E. Sørensen (1997), Multivessel batch distillation, *AIChE J.*, **43**, 971.

Stephanopoulos G. (1980), Synthesis of process flowsheets: An adventure in heuristic design or a utopia of mathematical programming, *Proceedings of International Conference on Foundations of Computer-Aided Chemical Process Design*, **2**, 439.

Exercises

6.1 Present the degrees of freedom analysis for the batch stripper.

6.2 Describe the degrees of freedom analysis for the middle vessel column.

6.3 Present the feasibility conditions for the batch stripper based on the shortcut method.

6.4 Present the feasibility conditions for the middle vessel column.

7

COMPLEX SYSTEMS

CONTENTS

The previous chapters except the last chapter described different aspects of batch rectification including development of a hierarchy of models ranging from simplified to rigorous, degrees-of-freedom analysis, and optimization. All along, the discussion was focused on batch rectification. The last chapter presented various complex configurations of batch columns. However, all along to bring out the basic concepts of batch distillation thermodynamically complex systems were eliminated from the discussions as far as possible. The following sections are devoted to distillation of thermodynamically complex systems like azeotropic, extractive, and reactive systems.

7.1 Azeotropic Systems

In a normal distillation column, the vapor becomes steadily richer in the more volatile component as it passes through successive plates. In azeotropic mixtures this steady increase in concentration does not take place, owing to the so-called azeotropic points. For instance, when a mixture of ethyl alcohol and water is distilled, the concentration of the alcohol steadily increases until it reaches 95.6 percent by weight, when the composition of the vapor equals that of the liquid, and no further enrichment occurs (see Figure 7.1). Such a mixture is called an azeotrope and cannot be separated by straightforward distillation.

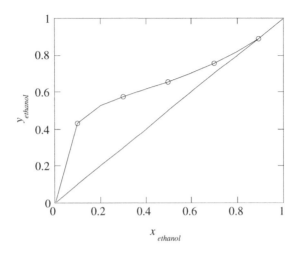

FIGURE 7.1
Vapor–Liquid Equilibrium Curve for the Ethanol–Water System

Azeotropic distillation is an important and widely used separation technique, as a large number of azeotropic mixtures are of great industrial importance. Despite their importance, azeotropic distillation techniques remain poorly understood from a design standpoint. This is because of the complex thermodynamic behavior of the system. Theoretical studies on azeotropic distillation have mainly centered around methods for predicting the vapor-liquid equilibrium data from liquid solution models and their application to distillation design (vanDongen and Doherty, 1985a). However, only during the past decade or two has there been a concerted effort to understand the nature of the composition region boundaries. Doherty and co-workers (Doherty and Perkins, 1977, 1978a, b, 1979; vanDongen and Doherty, 1985a, b; Foucher et al., 1991, Bernot et al., 1990, 1991, to mention a few) in their pioneering work proposed several new concepts in azeotropic distillation. They have established the use of ternary diagrams and residue curve maps in the design and synthesis of azeotropic continuous distillation columns. In batch distillation, they have outlined a synthesis procedure based on these residue curve maps. The residue curve maps were used by Kalagnanam and Diwekar (1993) to extract qualitative behavior based on which they extended the shortcut method to ternary azeotropic systems. A residue curve map is a map of liquid composition remaining in a simple distillation still over time. A residue curve map analysis involves terms such as saddle points, nodes, separatrix, etc. The following paragraph presents a mathematical basis for these terms.

7.1.1   Qualitative Analysis of Azeotropic Systems

The work by Doherty and co-workers focuses on analyzing the continuous column design and synthesizing from the qualitative behavior of the system. For example, the binary azeotropic curve shown in Figure 7.1 can be divided into two regions and a separate vapor-liquid-equilibria equation can be applied to each of the two regions. This enforces the impassable barriers of the azeotropic points, and feed composition lying in one region will always have distillate or bottom composition in the same region. Therefore, it can be easily said that the azeotropic points divide the equilibrium curve into two separate regions in which the behavior of the system is essentially similar. While it is easy to divide the regions of qualitative behavior of a binary system, the problem becomes more complicated for ternary and multicomponent systems. Doherty and co-workers used simple distillation curve maps to identify the qualitative behavior of ternary and quaternary systems. Figure 7.2 shows the residue curve map for the acetone-isopropanol-water system where the line AD divides the diagram into two qualitatively similar regions. It is possible to understand the residue curve map information from the qualitative analysis of the planar differential equation (Kalagnanam and Diwekar, 1993) given below.

Planar Differential Equation

$$\frac{dx}{dt} = P(x,y), \quad \frac{dy}{dt} = Q(x,y) \tag{7.1}$$

In the above equation, any state (x,y) is represented by a point in the *phase plane*, which represents the set of all possible states of the system. The state variables x and y are functions of time t: $x = f(t), y = g(t)$. As t varies, the state (x,y) moves along a *path* which is also called a *trajectory*. A complete path represents the history of the system throughout all times. The totality of all paths represents all possible histories and it is called the *phase portrait*. One and only one path passes through each point in the phase plane for autonomous systems such as the one we are dealing with here. So the precise knowledge of any single point on the path determines the entire path for the system.

Qualitative investigation entails the construction of a *phase portrait* which describes the partitioning of the phase plane into regions in which the behavior of the system is equivalent. Three kinds of information are critical for the construction of the phase portrait. First we need to identify the point trajectories or the *equilibrium points* of the system and the local behavior of the system around these points. The number of possible behaviors in the neighborhood of simple equilibrium points is finite and they are shown in Figure 7.3. The local behavior is derived by computing the eigenvalues and eigenvectors of the Jacobian of the system around the equilibrium points. The sign of the real parts determines the stability of the equilibrium point and the existence or nonexistence of the real and imaginary points determines the topological structure around the equilibrium point. Table 7.1 presents the criteria for determining

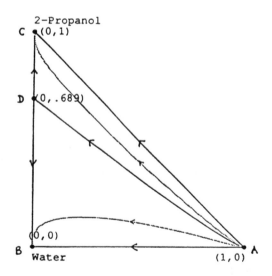

FIGURE 7.2
Qualitative Regions in the Residue Curve Map

the stability and topological structure using the eigenvalues and eigenvectors. We also need to identify the existence or the nonexistence of closed paths called *limit cycles* which are periodic solutions of the system. Last, we need to identify the global behavior of the trajectories that pass through equilibrium points. Such paths are called *separatrices*. In general, saddle points give rise to one stable and one unstable separatrix. Stable nodes give rise to stable separatrices and unstable nodes unstable separatrices. For a more complete discussion, the reader is referred to Kalagnanam (1991).

TABLE 7.1
Local Behavior of the Equilibrium Points

Classification	Criteria	Stability
Center	$Re(\lambda_1) = 0;\ Re(\lambda_2) = 0;$ $Im(\lambda_1) \neq 0;\ Im(\lambda_2) \neq 0$	Metastable
Node	$Re(\lambda_1) \leq 0;\ Re(\lambda_2) \leq 0;$ $Im(\lambda_1) = 0;\ Im(\lambda_2) = 0$	Stable
Node	$Re(\lambda_1) \geq 0;\ Re(\lambda_2) \geq 0;$ $Im(\lambda_1) = 0;\ Im(\lambda_2) = 0$	Unstable
Saddle Point	$Re(\lambda_1) \leq 0;\ Re(\lambda_2) \geq 0;$ or Vice Versa	Unstable
Focus	$Re(\lambda_1) \leq 0;\ Re(\lambda_2) \leq 0;$ $Im(\lambda_1) \neq 0;\ Im(\lambda_2) \neq 0$	Stable
Focus	$Re(\lambda_1) \geq 0;\ Re(\lambda_2) \geq 0;$ $Im(\lambda_1) \neq 0;\ Im(\lambda_2) \neq 0$	Unstable

Residue Curve Maps for Azeotropic Systems

Now we turn our attention to the thermodynamics of azeotropic distillation processes described using differential equations. Doherty and co-workers (van-Dongen and Doherty, 1985a; Bernot et al., 1990; Foucher et al., 1991) have shown the correspondence between the mathematical properties of the differential equations describing the simple distillation residue curve map and thermodynamic properties for azeotropic distillation.

In Chapter 1 we saw the simple distillation equation (Equation 1.1) in terms of distillate and bottom compositions. Doherty et al. extended the equation to derive a residue curve map for the simple distillation operation. The residue curve map graphs the liquid composition paths which are solutions to the following set of ordinary differential equations (vanDongen and Doherty, 1985a):

$$\frac{dx_i}{d\zeta} = x_i - y_i \qquad i = 1, 2, \ldots, n-1 \tag{7.2}$$

where n is the number of components in the system (e.g., for ternary systems

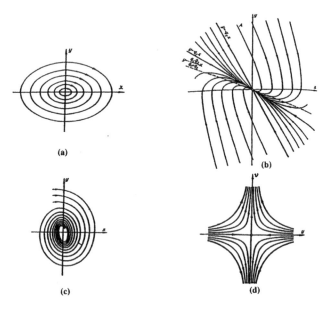

FIGURE 7.3
Local Behavior of the Equilibrium Points: (a) Center, (b) Node, (c) Focus, (d) Saddle Point

$n = 3$), the independent variable ζ is a monotonically increasing quantity related to real time, and

$$0 \leq x_i \leq 1, \; 0 \leq y_i \leq 1$$

The vapor composition y_i is related to x_i, the liquid composition, by the following vapor-liquid equilibrium (VLE) relationship

$$y_i = f(x_i, \; TE, P) \qquad i = 1, 2, \ldots, n$$

The relation between x_i and y_i is given by an implicit relation which involves temperature TE and pressure P. A residue curve map is a phase portrait of the system in the composition space. The azeotropes correspond to the equilibrium points of the system and the separatrices along with the boundaries of the phase portrait form the distillation boundaries.

The topological structure of the residue map is severely constrained by the thermodynamics of azeotropic distillation. Since x_i is the liquid composition, $\sum_i x_i = 1$ and the residue curve map for a ternary system is defined on a triangular diagram (simplex). The vertices correspond to pure component compositions and the equilibrium points are binary and ternary azeotropes. For $n = 3$ the residue curve map is defined on a right-angled or equilateral triangle. The temperature surface is a naturally occurring Lyapunov function for the above differential equations. The movement of the liquid composition x_i is always in a direction which makes the temperature increase. The temperature function can be used to show that the above differential equations do *not* possess limit cycles. This constraint is extremely useful in constructing the residue curve map since it eliminates the computationally intensive search for the existence or nonexistence of limit cycles. Moreover, it has also been shown that the equilibrium points of this system are isolated and are either *saddles* or *nodes* (Doherty and Perkins, 1978a). This information severely restricts the search for the global configuration of the separatrices. Foucher et al. (1991) provide an automatic procedure for the determination of the structure of simple distillation residue curve maps for ternary mixtures. They assume knowledge of the boiling points and compositions at the azeotropic points and use rules to classify the azeotropes as nodes or saddles. Bossen et al. (1993) present the computational tool needed for the simulation, design, and analysis of azeotropic distillation columns in general by simulating the separatrices directly. Kalagnanam and Diwekar (1993) used the mathematical definitions of saddles and nodes to obtain information about nature of the azeotropic (equilibrium) points and generate the separatrices, based on linearization information along specific directions. The following example illustrates their procedure for constructing a residue curve map and extracting regions of qualitatively similar behavior.

Example 7.1: The vapor-liquid equilibrium for the ethanol-water-cyclohexane system is given by the following relationships:

$$y_i = \frac{x_i \; \gamma_i \; p_i^s}{p} \qquad i = 1, 2, \ldots, n$$

where the liquid nonideality in terms of the activity coefficients γ can be calculated using the Wilson[1] equation given below.

$$\ln \gamma_i = -\ln \left(\sum_{j=1}^n x_j \Lambda_{ij} \right) + 1 - \sum_{k=1}^n \frac{x_k \Lambda_{ki}}{\sum_{j=1}^n x_j \Lambda_{kj}} \qquad i = 1, 2, \ldots, n$$

and the Antoine vapor pressure equation

$$\log \frac{P_i^s}{P} = A_i - \frac{B_i}{TE + C_i} \qquad i = 1, 2, \ldots, n$$

The temperature TE (in degrees Kelvin) in the above equations is calculated using

$$\sum_{i=1}^n \frac{x_i \, \gamma_i \, P_i^s}{P} = 1$$

and A_i, B_i, C_i are constants associated with the i-th component in the ternary system and Λ_{ij} is the interaction parameter between component i and component j given by

$$\Lambda_{ij} = \frac{V_j}{V_i} \exp \left(-\frac{(\lambda_{ij} - \lambda_{ii})}{R \, TE} \right)$$

V_i is the molar volume of pure liquid components i and λ_{ij} is the interaction energy between components i and j. R is the gas constant. The vector coefficients $\vec{A}, \vec{B}, \vec{C}, \vec{V}, \vec{\Lambda}$ for the ethanol-water-cyclohexane system are given in Table 7.2.

TABLE 7.2
System Parameters for the Ethanol–Water–Cyclohexane Ternary Azeotropic System

Component i	Antoine Constants			
	A_i	B_i	C_i	V_i
Ethanol	5.2314	1592.864	-46.816	58.68
Water	5.1905	1730.630	-39.574	18.07
Cyclohexane	3.9706	1206.470	-49.864	108.75

Component i	Wilson Constants		
	λ_{i1}	λ_{i2}	λ_{i3}
Ethanol	1.0	288.9156	1894.2908
Water	962.0073	1.0	18155.835
Cyclohexane	399.2968	35623.307	1.0

(a) Find the equilibrium points.

---

[1] Note that the system ethanol-water-cyclohexane forms a heterogeneous azeotrope (VLLE) and normally the Wilson equation cannot be used to predict the liquid-liquid phase split. However, it has been observed that if one does not differentiate between the two liquid phases then the Wilson equation can predict the average VLE behavior as seen here.

(b) Classify the equilibria.

(c) Find the configuration of the separatrices.

Solution: The complete residue curve map for the *ethanol-water-cyclohexane* system is provided in Figure 7.4 and is derived using the following steps:

(a) Finding the equilibria: At equilibrium, $dx_i/d\zeta = 0$. The implicit equations are solved to find the equilibrium (azeotropic points). This system has *three* binary azeotropes, $(0.457, 0)$, $(0, 0.299)$, and $(0.89, 0.11)$. The system has *one* ternary azeotrope, $(0.316, 0.15)$.

(b) Classify the equilibria: The classification of the equilibrium points is based on the local behavior of the system about the equilibrium points. In order to classify the equilibrium points, we need to linearize the equations about the equilibrium points and evaluate the eigenvalues and eigenvectors of the linear system. For a ternary system this procedure is straightforward and the partial derivatives are calculated using perturbation methods. However, for equilibrium points on the boundaries as in the case of binary azeotropes and the vertices we need to remember to evaluate the partial along directions from within the triangle and then use these partials to evaluate the eigenvalues and eigenvectors. We will use simple rules regarding the appropriate directions for evaluating the partials for the vertices and binary azeotropes. For the *ethanol-water-cyclohexane* system, all the binary azeotropes are *saddles* $(Re(\lambda_1) \leq 0; \ Re(\lambda_2) \geq 0)$ and all the vertices are *stable nodes* $(Re(\lambda_1) \leq 0; \ Re(\lambda_2) \leq 0; \ Im(\lambda_1) = 0; \ Im(\lambda_2) = 0)$. The ternary azeotrope is an *unstable node* $(Re(\lambda_1) \geq 0; \ Re(\lambda_2) \geq 0; \ Im(\lambda_1) = 0; \ Im(\lambda_2) = 0)$.

(c) Configure the separatrices: Once we have all the equilibrium points of a given system we can list all the separatrices of the system, since we know that saddle points give rise to *one* stable and *one* unstable separatrix and stable (unstable) nodes give rise *two* stable (unstable) separatrices. We also have available to us the local behavior of these separatrices. The characterization of the global qualitative behavior of the separatrices is just a question of how to link up each separatrix beginning at one equilibrium point with its corresponding end at another (or the same) point. The search for the possible global configuration is constrained by rules derived from topological considerations. Here we discuss how these rules constrain the separatrices for the *ethanol-water-cyclohexane* system. Since all the binary azeotropes are saddle points and lie on the edges of the triangle, the edges are necessarily the separatrices of the binary azeotropes. Moreover, all the vertices are stable nodes; therefore, the unstable separatrices emanating from each of the binary azeotropes terminate at the vertices moving along the edges. Referring to Figure 7.4, it may be seen that the unstable separatrices for each of the binary azeotropes emanate from the unstable ternary azeotrope and divide the residue curve map into *three* qualitatively different regions. The unstable separatrices of the ternary azeotrope asymptotically evolve toward one of the vertices.

The above example demonstrated how information about composition boundaries can be obtained from residue curve maps. This qualitative information is very important for design and synthesis of columns. Diwekar and

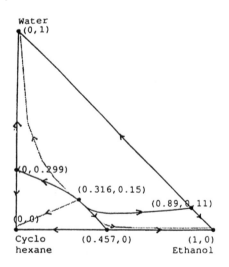

FIGURE 7.4
Residue Curve Map

co-workers (Diwekar, 1991; Kalagnanam and Diwekar, 1993) used these distillation boundary regions to extend the shortcut method described in Chapter 3 to azeotropic systems, as presented below.

7.1.2   Extending the Shortcut Method to Azeotropic Systems

The shortcut methods presented in Chapter 3 are based on the assumption of constant relative volatility throughout the column, updated at each time instant, and they use the following relationship for obtaining vapor-liquid equilibria:

$$y_i = \frac{\alpha_i x_i}{\sum_{k=1}^{n} \alpha_k x_k}; \quad i = 1, 2, \ldots, n \tag{7.3}$$

where y_i is the vapor composition of component i in equilibrium with the liquid composition x_i of component i and α_i is the relative volatility of component i.

However, in the case of azeotropic systems this relation is no longer valid because of the azeotropic points (where the relative volatility becomes unity) and the distillation boundaries (which bound the distillation paths). The azeotropic points and distillation boundaries offer an impassable barrier or barriers (refer to Figures 7.1 and 7.4). For binary azeotropic systems, Anderson and Doherty (1984) transformed the variables of binary vapor-liquid equilibria calculations by splitting the equilibrium curve into two regions. This approach is used for extending the shortcut method to binary azeotropic systems (Diwekar, 1991). The bottom curve below the azeotropic composition (Figure 7.5) of component 1 is represented by

$$x_1' = \frac{x_1}{x_1^{az}}; \quad x_2' = 1 - x_1'$$
$$y_1' = \frac{y_1}{y_1^{az}}; \quad y_2' = 1 - y_1' \tag{7.4}$$

and the top curve is given by

$$x_1' = \frac{x_1 - x_1^{az}}{1 - x_1^{az}}; \quad x_2' = 1 - x_1'$$
$$y_1' = \frac{y_1 - y_1^{az}}{1 - y_1^{az}}; \quad y_2' = 1 - y_1' \tag{7.5}$$

The equilibrium relationship for the binary system in terms of the transformed variables is represented by

$$y_1' = \frac{\alpha_1 x_1'}{1 + (\alpha_1 - 1)x_1'} \tag{7.6}$$

Table 7.3 presents this extended shortcut model for binary azeotropic systems, and the simulation model for all the three modes of operation is shown in Figure 7.6.

TABLE 7.3
Modified Model Equations for Binary Azeotropic Systems

Variable Reflux	Constant Reflux	Optimal Reflux
Differential Material Balance Equation $$x^{(i)}_{B_{new}} = x^{(i)}_{B_{old}} + \frac{\Delta x^{(1)}_B\left(x^{(i)}_D - x^{(i)}_B\right)_{old}}{\left(x^{(1)}_D - x^{(1)}_B\right)_{old}}, i = 1,2,\ldots,n$$		
Variable Transformation *bottom column* $$x^{(i)}_B = \frac{x^{(i)}_B}{x^{az}_i}, i = 1,2,\ldots,n$$ *top column* $$x^{(i)}_B = \frac{x^{(i)}_B - x^{az}_i}{1 - x^{az}_i}, i = 1,2,\ldots,n$$		
Hengestebeck–Geddes Equation $$x^{(i)}_D = \left(\frac{\alpha_i}{\alpha_1}\right)^{C_1}\frac{x^{(1)}_D}{x^{(1)}_B}x^{(i)}_B, i = 2,3,\ldots,n$$ Unknowns		
R, C_1	$C_1, x^{(1)}_D$	$R, C_1, x^{(1)}_D$
Summation of Fractions $$\sum_{i=1}^n x^{(i)}_D = 1$$		
C_1 Estimation $$\sum_{i=1}^n \left(\frac{\alpha_i}{\alpha_1}\right)^{C_1}\frac{x^{(1)}_D}{x^{(1)}_B}x^{(i)}_B = 1$$	*$x^{(1)}_D$ estimation* $$x^{(1)}_D = \frac{1}{\sum_{i=1}^n\left(\frac{\alpha_i}{\alpha_1}\right)^{C_1}\frac{x^{(i)}_B}{x^{(1)}_B}}$$	
Fenske Equation $Nmin \approx C_1$ *Underwood Equations* $$\sum_{i=1}^n \frac{\alpha_i x^{(i)}_B}{\alpha_i - \phi} = 0 \ ; \ Rmin\ u + 1 = \sum_{i=1}^n \frac{\alpha_i x^{(i)}_D}{\alpha_i - \phi}$$ *Gilliland's Correlation*		
R Estimation $R = F(N, Nmin, Rmin\ u)$	*C_1 Estimation* $Rmin\ g = F(N, Nmin, R)$ $\frac{Rmin\ g}{R} - \frac{Rmin\ u}{R} = 0$	
		R Estimation $R = F(Min.\ H)$
Variable Retransformation *bottom column* $$x^{(i)}_B = x^{(i)}_B x^{az}_i, i = 1,2,\ldots,n$$ $$x^{(i)}_D = x^{(i)}_D x^{az}_i, i = 1,2,\ldots,n$$ *top column* $$x^{(i)}_B = x^{(i)}_B - x^{az}_i(1 - x^{(i)}_B), i = 1,2,\ldots,n$$ $$x^{(i)}_D = x^{(i)}_D - x^{az}_i(1 - x^{(i)}_D), i = 1,2,\ldots,n$$		

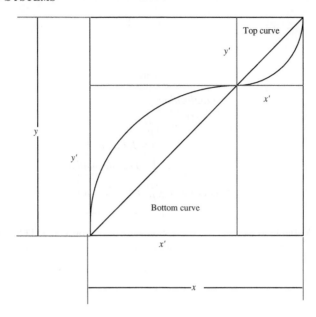

FIGURE 7.5
Variable Transformation for Binary Azeotropic Systems

Similar to the original shortcut method, at each instant the bottom compositions of all the components are obtained from the differential material balance equations. Then the still compositions, $x_B^{(i)}, i = 1, 2, \ldots, n$, are transformed in terms of the azeotropic points using the above-mentioned transformation formulae (Equations 7.4-7.5). For the variable reflux operation, the distillate composition of the key component $x_D^{(k)}$ is also transformed. These transformed variables are then used in the calculation of distillate compositions of all the components, $x_D^{(i)}, i = 1, 2, \ldots, n$, the Hengestebeck–Geddes constant C_1, and the reflux ratio R (for the variable reflux mode). The initial variable transformation essentially results in transformed distillate compositions, which need to be retransformed to obtain the actual values.

It has been found that preliminary estimates of this extended shortcut method for ideal binary azeotropic systems compare very well with the rigorous models and the extended model offers significant computational savings over the rigorous models (for some cases, the shortcut model was found to be 200-300 times faster than the rigorous models). For details of these preliminary studies please refer to Diwekar (1991). A few of these cases for the input data given in Table 7.4 are shown in Figure 7.7.

For ternary systems, although the transformation of variables is critical for extension of the shortcut method, the impassable barriers are not represented by the azeotropic points as in the case of binary azeotropic batch distillation. Instead, the distillation boundaries play the important role of representing

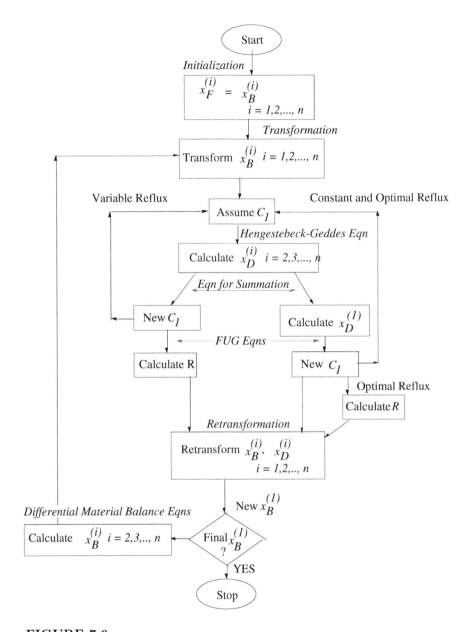

FIGURE 7.6
Flowchart for the Extended Shortcut Method for Binary Azeotropic Systems

TABLE 7.4
Input Data for the Test Problems

Problem	System	N	R	V	Feed Comp.	
1	Ethyl Acetate/Ethanol	8	2.73	12	0.2	0.8
2	Chloroform/Methanol	8	5.0	20	0.2	0.8
3	Methanol/Ethyl Acetate	8	2.73	12	0.2	0.8
4	Toluene/n-Butanol	8	2.73	12	0.2	0.8

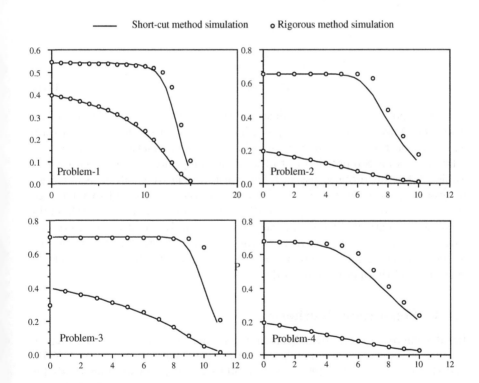

FIGURE 7.7
Transient Distillate Composition Profiles

the impassable barriers. These distillation boundaries can be identified from residue curve maps for simple distillation operations. Figure 7.2 shows such a residue curve map for an isopropanol-water-acetone system. Following the argument from continuous distillation theory, Kalagnanam and Diwekar (1993) proposed that the distillation boundaries divide the ternary diagrams into different qualitative regions. The variables in each region are transformed in such a way that each region will form a separate triangular diagram. The thermodynamics of each region is represented by the constant relative volatility equations. The shortcut method and the modified shortcut method are then extended using these "transformed variable" thermodynamic models.

To illustrate the procedure we will consider an example of a ternary system containing acetone-isopropanol-water as given by Kalagnanam and Diwekar (1993). This system has one binary azeotrope and one separatrix. As a first approximation this separatrix is approximated by a straight line. This results in two regions, ABD and BDC (Figure 7.2).

In Figure 7.2, the triangle ABD is represented by the following transformed variables:

$$
\begin{aligned}
x_1' &= x_1; \quad y_1' = y_1 \\
x_2' &= \frac{x_2}{x_2^{az}}; \quad y_2' = \frac{y_2}{y_2^{az}} \\
x_3' &= 1 - x_1' - x_2'; \quad y_3' = 1 - y_1' - y_2'
\end{aligned}
\tag{7.7}
$$

The vapor-liquid equilibria data for the three components is generated using the Wilson equation for liquid phase nonidealities. The vapor pressure equation constants and the constants in the Wilson equation needed to correlate the vapor composition y_i to the liquid composition x_i are presented in Table 7.5 for this ternary system.[2] The constant relative volatility is obtained by averaging over the triangle ABD and using the transformed variables. To check the thermodynamic approximation, the rigorous model for batch distillation is also extended to include the constant relative volatility transformed variable model. The results of both the extended shortcut method and the extended rigorous method are successfully compared to the rigorous method with the real thermodynamic model.

7.1.3 Separation Synthesis

Residue curve maps occupy a significant place in the conceptual design stage of distillation sequencing. For batch distillation, vanDongen and Doherty (1985a) and Bernot et al. (1990, 1991) tried to address this problem of synthesis for ternary systems and then extend it to quaternary systems. The following analysis is based on these papers. The batch distillation synthesis problem

---

[2] In the Antoine equation the pressure is in atmospheres and the temperature is in Kelvin, and in the Wilson equation constants are given in cal/mole. Data obtained from Gmehling and Onken (1982).

TABLE 7.5

System Parameters for the Acetone–Isopropanol–Water Ternary Azeotropic System and Input Parameters for the Batch Distillation Columns (Figure 7.2)

	System Parameters for Acetone-Isopropanol-Water							
Component	Antoine Constants			V_i	Wilson Constants			Rel. Vol. Model
	A_i	B_i	C_i		λ_{i1}	λ_{i2}	λ_{i3}	
Acetone	4.23633	1210.595	-43.336	74.05	1.0	-203.114	679.487	2.790
Isopropanol	5.99476	2010.330	-20.364	76.92	593.183	1.0	556.390	1.000
Water	5.19050	1730.630	-39.574	18.07	1251.917	1294.715	1.0	0.367

Input Parameters for the Batch Distillation Column

Component	Case (a)	Case (b)
Feed Composition		
Acetone	0.05	0.10
Isopropanol	0.40	0.40
Water	0.55	0.50
Number of Plates	5	5
Molar Vapor Rate	5.0	5.0
Molar Feed	100.0	100.0
Holdup %	6.0	0.6

involves decisions such as different cuts or fractions to be produced and the intermediate fractions to be recycled.

A single batch distillation is characterized by many trajectories: one for the still composition, one for the distillate, and one for each tray. The trajectory of the still composition is called a *still path* and the trajectory for the distillate composition a *distillate path*. The trajectories of the trays have the same features as the trajectory for the still (Bernot et al., 1990). For the case of maximum separation, the position of the still paths can be found from only a knowledge of residue curve maps. For large reflux ratios, the still paths straighten as the number of trays increases, following a direction of the unstable node through the initial composition, as shown in Figure 7.8 for the methanol-acetone-chloroform system. Each still path follows this straight line until it reaches either the edge of the triangle or a simple distillation stable separatrix. The path then follows the limiting curves up to a stable node. In short, the still boundaries correspond to the edges of the triangle and to the stable simple distillation separatrices, with a slight exception when these stable separatrices are slightly curved. To understand the distillate path we will first consider the residue curve equation given below.

$$\frac{dx_i}{d\zeta} = x_i - y_i \qquad i = 1, 2, \ldots, n-1 \qquad (7.8)$$

From the above equation it can be seen that at any given point on the residue curve map (please see Figure 7.9) x_i, the vector joining the liquid composition and the equilibrium vapor composition, is tangent to the residue curve map at that point. Each residue curve has a vapor boil-off curve representing the equilibrium composition y_i where the tangent to the residue curve intersects. For each batch distillation residue curve there will be a corresponding distillate curve, which denotes the locus of the distillate composition x_D as x_{DS} change with time during the course of the distillation. The residue curve map essentially represents the Rayleigh equation and as we have seen in the earlier chapters, the Rayleigh equation of simple distillation is valid for batch distillation with zero holdup plate columns. Therefore, the relationship of the distillation curve to the residue curve is the same as the relationship of the vapor boil-off curve to the liquid composition, and hence the tangency condition applies.

In general, if the mixture is not wide boiling, the total reflux curve can be approximated by the residue curve. Therefore, for a large number of plates and for a large reflux ratio, the liquid composition profile (distillation curve) in the column follows the residue curve in the reverse direction. Thus, the intersection of the tangent to the still path with the simple distillation residue curve map, i.e., the distillate composition, is on a curve that approaches these boundaries asymptotically as the number of trays increases. Therefore, the distillate paths are bounded by the edges of the triangle and the unstable separatrices of the simple distillation curve map. From the distillate path and

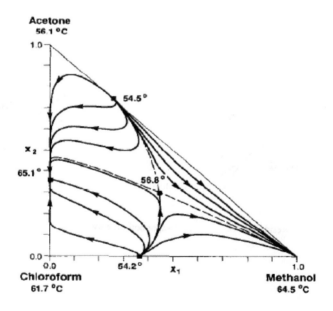

FIGURE 7.8
Simple Distillation Residue Curve Map for the Methanol–Acetone–Chloroform
System (Reproduced from Bernot et al., 1990)

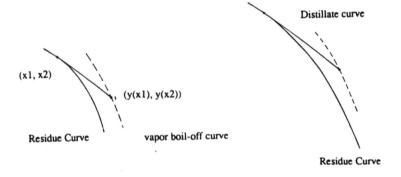

FIGURE 7.9
Residue Curve and Tangency Condition

still path, it is possible to predict the limiting fractions or cuts of the batch distillation.

The distillate paths and still paths are feed composition dependent. However, Bernot et al. (1990) defined a concept called *batch distillation region* where any initial condition taken in that region leads to the same sequence of cuts[3]. They defined the batch distillation region by the boundaries of the region for a ternary mixture. They provided the following rules to identify the batch distillation regions.

- If one stable separatrix or two (rarely three) adjoining separatrices divide the triangle into regions each containing an unstable node, this separatrix (or these separatrices) constitutes a batch distillation boundary.

- In each part of the triangle that contains an unstable node the other batch distillation boundaries are the straight lines linking the unstable node with 1) the saddles that are connected to the unstable node by an unstable separatrix and 2) the stable nodes.

- A straight line boundary should not intersect a stable separatrix.

- If the stable separatrix is highly curved, the straight line boundary is tangent to the stable separatrix.

The following example will illustrate this synthesis procedure for batch distillation of an acetone-methanol-chloroform mixture.

Example 7.2: The residue curve map for the methanol–chloroform–acetone system is shown in Figure 7.8.

a) Summarize the main features of the residue curve map.

b) Find the batch distillation regions for this system.

c) Define the cuts for each region.

Solution:

a) The main features of the residue curve maps are given below. The system has two low-boiling azeotropes, one high-boiling azeotrope, and a ternary saddle point. The saddle azeotrope has four separatrices entering and leaving it, which divide the map into two distinct simple-distillation regions. In the left hand region, all the residue curves are attracted to the high-boiling acetone-chloroform azeotrope. In the the right hand region, they are attracted to the pure methanol vertex. The two regions each contain one stable separatrix.

b) The boundary of the batch distillation region is divided by the stable separatrix which divides the diagram into two batch distillation regions, as shown in Figure 7.10a, and which gives as a first fraction the methanol-chloroform azeotrope. In each of these two parts, there is another boundary which links an unstable node to the ternary azeotrope through a straight path. This boundary defines the last fraction which can be either an acetone-chloroform azeotrope or pure

---

[3]This was first seen by Ewell and Welch (1945).

methanol. Since the distillate path has two possible routes, the regions shown in Figure 7.10a can be further subdivided into two regions, one corresponding to the path following the stable separatrix, and the other the path not following the stable separatrix. Figure 7.10b shows all the six batch distillation regions.

c) The fractions corresponding to each region are given in Table 7.6.

TABLE 7.6

Sequences of Fractions for the Methanol–Acetone–Chloroform Batch Distillation

Region	No. of Fractions	Sequence
1	3	Methanol-Chloroform Chloroform Acetone-Chloroform
2	5	Methanol-Chloroform Ternary Azeotrope Methanol-Acetone Acetone Acetone-Chloroform
3	4	Methanol-Chloroform Ternary Azeotrope Methanol-Acetone Methanol
4	4	Methanol-Acetone Ternary Azeotrope Methanol-Acetone Methanol
5	5	Methanol-Acetone Ternary Azeotrope Methanol-Acetone Acetone Acetone-Chloroform
6	3	Methanol-Acetone Acetone Acetone-Chloroform

Ahmad and Barton (1996) and Ahmad et al.(1998) used nonlinear dynamics and topology concepts to set the rigorous framework that enables us to find the batch distillation boundaries for multicomponent systems. They were motivated by the failures noticed with the methodology proposed by Bernot et al., (1990) which cannot be applied to ternary diagrams that involve a stable ternary node and a single unstable node. In these cases, the stable separatrix that constrains the still path does not divide the composition space. The same problem was pointed out by Safrit and Westerberg (1997a), who introduced specific rules to determine basic distillation boundaries. Rodriguez-Donis et

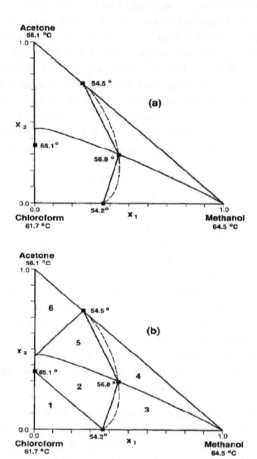

FIGURE 7.10
Batch Distillation Regions for the Methanol–Acetone–Chloroform System

al.(2001a) presented both the Ahmad et al., (1998) and the Safrit and Westerberg (1997a) approaches and highlighted the similarities, which are given
below.

Under the assumption of straight distillation boundaries, Safrit and Westerberg (1997a) established an algorithm to generate basic and batch distillation regions for azeotropic mixtures with k components. A basic distillation
region is defined as a set of residue curves that start at a specific unstable node
and terminate at a particular stable node. First, they say that if there is a
k-dimensional unstable (stable) azeotrope, any path of decreasing (increasing)
temperature from a stable (unstable) fixed point that ends in a saddle must be
supplemented by the connection of the azeotrope to the saddle, this connection being a stable (unstable) separatrix. After this preprocessing step, they
defined a maximum (minimum) residue surface, denoted MaxRS (MinRS), as
a manifold that separates a composition space into subregions, each containing one unstable (stable) node. The MaxRS (MinRS) is the intersection of
the lists of fixed points that can be reached along basic boundaries and edges
of the ternary diagram with increasing (decreasing) temperature from each
unstable (stable) node up to each stable (unstable) node. Hence, the maximum (minimum) residue surface defines the stable (unstable) separatrices of
the ternary system. The combination of the maximum residue surface and
the minimum residue surface determines the basic distillation regions. Basic
distillation regions are then divided into one or more batch distillation regions
depending on the batch column configuration. For the rectifier (stripper) configuration, batch distillation region boundaries are found by connecting each
unstable (stable) node of the basic region to every other fixed point. Batch
distillation boundaries for the rectifier and stripper configurations might not
be identical. However, connections between unstable and stable nodes in basic regions are the same for both rectifier and stripper configurations if basic
distillation boundaries are straight

The algorithm established by Ahmad et al. (1998) demonstrates that the
composition simplex for k-component mixtures can be obtained by determining the complete unstable or stable boundary limit sets of each fixed point.
New concepts such as the unstable (stable) manifold, stable (unstable) dividing boundary, and unstable (stable) boundary limit set of a fixed point are
introduced. The unstable manifold is defined as the region that contains all
simple distillation trajectories approaching a fixed point with decreasing time.
Hence, the unstable manifold of an unstable node has a dimension equal to
$k - 1$. If the system contains two or more unstable nodes, unstable manifolds
will be separated with a $(k - 2)$-dimensional surface called a stable dividing
boundary. These surfaces are similar to the maximum residue surface defined
by Safrit and Westerberg (1997a). Likewise, the unstable dividing surface is
equivalent to the minimum residue surface. The stable dividing boundary is
determined from the common unstable boundary limit set, which is obtained
by defining an unstable boundary limit set for each unstable node. The un-

stable boundary limit set of the unstable node involves all fixed points that are limit points to its unstable manifold, except for the unstable node itself.

However, Ahmad et al. (1998) presented a more complete algorithm to define the batch distillation product cut sequence in the case of multicomponent mixtures. Considering linear pot composition boundaries, the product cut sequence is enumerated from the adjacency matrix A_k using a formulation of the problem as a graph theoretical problem. This formulation provides a set D of k points that starts at a batch distillation region unstable node and contains the remaining batch distillation region fixed points in increasing order of boiling temperature. The authors defined four categories to describe all possible characteristics of D that can be found in ternary diagrams. They established necessary and sufficient conditions for determining the k feasible product simplex P depending on the D category. They showed that the product simplex P formed by the k fixed points can either coincide with or be a subset of the respective batch distillation region. Moreover, these conditions permit the cases in which the batch distillation region determined from the adjacency matrix provides an infeasible product sequence using batch distillation to be discarded.

Rodriguez-Donis et al. (2001a) completed the analysis of Ahmad et al. (1998) and argued that systems with a ternary azeotrope can also exhibit such behavior, provided that the unstable (stable) node (pure component or binary azeotrope) is linked to another fixed point included in the same batch distillation region. In these cases, the unstable (stable) separatrix might intercept a stable (unstable) separatrix depending on the location of the ternary azeotrope in the triangular diagram. It is then possible to obtain a batch distillation region whose product simplex P cannot be considered as a feasible product sequence, because there is no initial composition point that gives rise to this product sequence. Although Ahmad et al. (1998) studied only a batch rectifier configuration, the methodology presented can include a set of specific rules associated with other alternative batch distillation technologies.

Safrit and Westerberg (1997a) used their own methodology for finding the batch distillation regions to assess the synthesis of batch column sequences needed to separate the binary mixture into its pure components. These sequences depend on the initial feed composition, which is represented using the vector system developed earlier by Ahmad and Barton (1996). The batch column network is then determined using a state-task network, in which the states are mixtures and the tasks are various distillation column configurations applied to the states. Under the assumptions of total reflux/reboil ratio and an infinite number of trays, Safrit and Westerberg (1997b) considered four different batch column configurations: rectifier, stripper, middle vessel column, and extractive middle vessel column as distillation tasks. A task sequence is a combination of those four batch column configurations. The state-task network algorithm was tested on the ternary mixture acetone-chloroform-methanol.

Analysis of the azeotropic distillation in the middle vessel column was presented by Cheong and Barton (1999). The assumptions underlying this

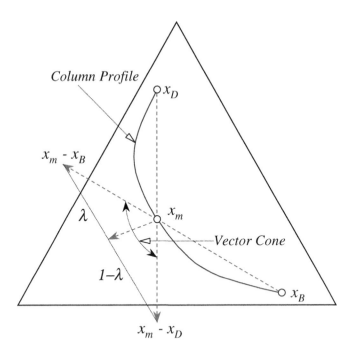

FIGURE 7.11
One way of Steering the Middle Vessel Composition Where λ is a Dimensionless Middle Vessel Parameter (Reproduced from Cheong and Barton, 1999)

analysis were an infinite number of trays, infinite reflux and infinite reboil ratios, and linear separation boundaries based on the warp time analysis. The top and bottom products are governed by the concept of steering the middle vessel composition (x_m), which is the vector cone of possible motion for the middle vessel composition. The vector is analyzed in terms of the dimensionless *middle vessel parameter* (λ), where

$$\lambda = \frac{D}{D + B} \tag{7.9}$$

This λ is related to q' as defined earlier. By varying λ (i.e., changing the production rates), it is possible to cross the middle vessel boundaries and thus overcome distillation barriers. One way of steering the middle vessel composition using λ is shown in Figure 7.11.

7.2 Extractive Distillation

Batch distillation of nonideal mixtures usually produces the azeotropes often associated with those mixtures. Among various techniques available to break azeotropes, extractive distillation processes have a place of choice. They rely upon the addition of a suitable entrainer to the mixture. Entrainer screening is therefore a key step for the synthesis and design of these processes[4]. Rodriguez-Donis et al. (2001a, b) devised a complete set of rules for the selection of a suitable entrainer from an analysis of all ternary residue curve maps under assumptions of a large number of stages and total reflux/reboil ratio. These rules are presented below in the next two subsections.

7.2.1 Entrainer Selection Rules for Homogeneous Batch Distillation

Screening rules are defined from the analysis of the ternary diagrams that enable a feasible batch distillation process. For simplicity straight line distillation boundaries are considered. The binary mixture consists of components A, B, and entrainer E. The structure of a particular batch task sequence depends on the residue curve map topology for the ternary mixture formed by the addition of entrainer to the binary feed mixture. Two general approaches can be used for selecting an entrainer in homogeneous batch distillation: (1) selection of an entrainer such that all of the desired products lie within the same batch distillation region and (2) selection of an entrainer such that the desired products lie in different batch distillation regions and the curvature of distillation boundaries is exploited. These approaches can be used to define four general conditions for selecting candidate entrainers that can be used for the separation of minimum-boiling azeotropic binary mixtures, maximum-boiling azeotropic binary mixtures, and close-boiling binary mixtures by homogeneous batch distillation. These conditions are given below.

1. No additional azeotrope is introduced.
2. A minimum-temperature binary azeotrope exists between E and either A or B.
3. A maximum-temperature binary azeotrope exists between E and either A or B.
4. There might be a ternary azeotrope.

Tables 7.7-7.9 summarize all of the conditions that a suitable entrainer must fulfill to enable feasibility of homogeneous batch distillation for the separation of minimum-temperature azeotropic binary mixtures, maximum-temperature

---

[4]Batch extractive distillation is also referred to as BED.

TABLE 7.7
Selection of Entrainers Enabling the Separation of Minimum-Boiling-Temperature Azeotropic Binary Mixtures by Homogeneous Batch Distillation. Straight Boundaries Case (Rodriguez-Donis et al., 2001a)

	Entrainer		
	Low Boiling	Intermediate Boiling	High Boiling
1		1	
2		$4[un]$	
3		$3(A)[s]$	
4		$2(B) + 3(A)[s]$	
5		$3(A) + 4[un]$	
6	$2(B)[un \leftrightarrow s] + 3(A)[s] + 4[un]$		

TABLE 7.8
Selection of Entrainers Enabling the Separation of Maximum-Boiling-Temperature Azeotropic Binary Mixtures by Homogeneous Batch Distillation. Straight Boundaries Case (Rodriguez-Donis et al., 2001a)

	Entrainer		
	Low Boiling	Intermediate Boiling	High Boiling
1		1	
2		$4[sn]$	
3			$2(B)[s]$
4			$2(B)[s] + 3(A)$
5			$2(B)[s] + 4[sn]$
6		$2(B) + 3(A)[sn \leftrightarrow s] + 4[sn]$	

azeotropic binary mixtures, and close-boiling binary mixtures, respectively. The rules are established according to the following code:

condition 1, 2, 3, or 4 (component A or B [stability un, sn, or s])

For example, condition 2 in Table 7.7, $4[un]$, can be read as condition 4 applies so a ternary azeotrope is present as an unstable node. A, B (Table 7.9) means that the rule applies to both A and B. The symbol $A \leftrightarrow B$ means that the condition can be found for A or B. Both commas and double arrows can apply to stability as well as to components.

The equation of topological indexes for ternary systems (Doherty and Perkins, 1979) can be used for the determination of the stability of the fixed points that have not been defined in these tables, as given below.

$$2N_3 - 2S_3 + N_2 - S_2 + N_1 = 2 \tag{7.10}$$

TABLE 7.9
Selection of Entrainers Enabling the Separation of Close-Boiling-Temperature Azeotropic Binary Mixtures by Homogeneous Batch Distillation. Straight Boundaries Case (Rodriguez-Donis et al., 2001a)

	Low Boiling	Entrainer Intermediate Boiling	High Boiling
1		1	
2		$2(A[un \leftrightarrow s], B)$	
3		$2(A, B) + 4[un \leftrightarrow s]$	
4		$3(A[sn \leftrightarrow s], B)$	
5		$3(A, B) + 4[sn \leftrightarrow s]$	
6	$3(A) + 4[s]$	$3(A) + 4[s]$ or $3(B) + 4[sn]$	$3(A \leftrightarrow B) + 4[sn]$
7	$2(A \leftrightarrow B)[un]$	$2(A[un] \leftrightarrow B[s])$	$2(B[s])$
8	$2(A \leftrightarrow B)4[un]$	$2(A) + 4[un]$ or $2(B) + 4[s]$	$2(B) + 4[s]$
9	$2(B) + 3(A[s])$	$2(B[un \leftrightarrow s] + 3(A[s \leftrightarrow sn])$	$2(B) + 3(A[sn])$
10	$3(A)$	$3(A \leftrightarrow B)$	
11	$2(B) + 3(A) + 4[un \leftrightarrow s]$	$2(B) + 3(A) + 4[sn \leftrightarrow un \leftrightarrow s]$	$2(B) + 3(A) + 4[sn \leftrightarrow s]$

where N and S stand for node and saddle, respectively. The subscript numbers 1, 2, and 3 refer to pure components, binary mixtures, and ternary mixtures, respectively.

As an application example, the rule defined as $2(B) + 3(A) + 4[un]$ (Case 6, Table 7.7) means that the entrainer must form a minimum-boiling-temperature azeotrope with component B, a maximum-boiling-temperature azeotrope with component A, and a minimum-boiling-temperature ternary azeotrope. Application of the stability equation (Equation 7.10) says that both binary azeotropes are saddles. For an entrainer with a boiling point intermediate between those of the binary components, likewise, the rule $2(A, B)$ describes a ternary mixture in which the entrainer forms a minimum-boiling azeotrope with each component. The rule $2(A \leftrightarrow B)$ describes two unlike ternary diagrams in which the entrainer introduces a minimum-boiling azeotrope with either A or B. If two ternary diagrams exhibit two maximum-boiling azeotropes between the entrainer and both components and if the difference between their topological structures depends on the stability of the fixed points, then the rule is defined as $3(A[sn \leftrightarrow s], B[s \leftrightarrow sn])$ and includes some stability information.

These rules along with the residue curve map are used to generate batch distillation product sequences. For example, to separate a minimum-boiling-temperature azeotropic mixture by a homogeneous azeotrope, using rule 1 in Table 7.7 acetone can be selected as an intermediate boiling entrainer. Figure 7.12 shows the ternary residue curve map for this system. The ternary diagram exhibits a single batch distillation region. The ethanol vertex is a stable node and can be obtained as a bottom product P_1 so the batch stripper can be used for this operation. Similarly, products P_3 and P_4 can be obtained in the second fraction of the batch stripper as shown in the figure.

Figure 7.13 shows an example for the separation of a water-acetic acid mixture with 1-propanol, a low-boiling entrainer selected from rule 7 of Table 7.9 for a close-boiling mixture. The acetic acid vertex is a stable node, so acetic acid can be obtained as the bottom product P_1 using a batch stripper process. When acetic acid is removed from the column, the second batch stripper task provides water as the bottom product P_3 until the composition of the 1-propanol-water azeotropic mixture P_4 is achieved in the reflux drum and is recycled to a further batch cycle to be added to a water-acetic acid mixture to give feed F.

Curved Distillation Boundaries

It has been shown by Bernot et al. (1990, 1991) that curved boundaries affect the product cut sequences of batch distillation. The still (reflux drum) path follows the stable (unstable) separatrix for the batch rectifier (stripper). When the still or reflux drum path moves on a curved separatrix from the concave side of the separatrix, the column profile switches into another region to follow the residue curves, and the product compositions are defined by the fixed

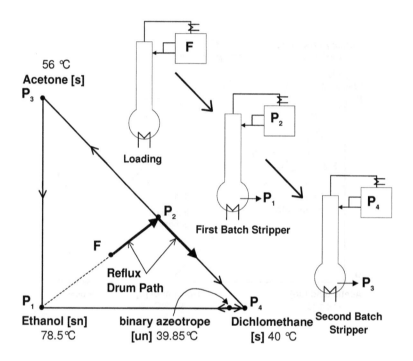

FIGURE 7.12
Example of Separation of a Minimum-Boiling Azeotrope Using an Intermediate Boiling Entrainer in a Homogeneous Batch Column (Rodriguez-Donis et al., 2001a)

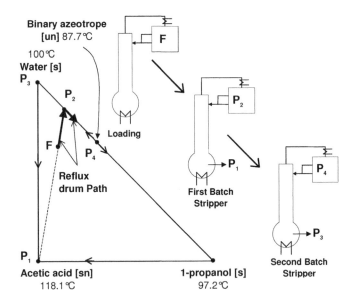

FIGURE 7.13
Example of Separation of a Close-Boiling Mixture Using a Low Boiling Entrainer in a Homogeneous Batch Column (Rodriguez-Donis et al., 2001a)

points of the new region. In particular, the curvature of the distillation boundaries is responsible for the peculiar behavior obtained experimentally and by
simulation in batch rectification where the distillate temperature decreases for
a while and then increases again, as seen in the straight boundaries. Therefore, the rules change for curved distillation boundaries. Tables 7.10 and 7.11
present the rules for residue curve maps for a curved distillation boundary.

Figure 7.14 shows the separation of a maximum-boiling temperature
azeotrope by homogeneous batch distillation, exploitation of the curvature
of distillation boundary. Here, benzene is the heavy entrainer used for separation of methyl acetate-choloroform. Benzene obeys the first rule in Table
7.11. Figure 7.14 shows the sequence of three batch tasks.

These rules can be used in Computer Aided Molecular Design (CAMD)
to obtain the best possible entrainer for a particular separation and then
residue curve maps can be used to obtain distillation task sequence. Kim et
al. (2004) used the rules and CAMD to obtain the best possible entrainers
for an industrial process for separation of acetonitrile and water. Figure 7.15
shows the ternary diagram which uses the low boiling entrainer propyl amine
obtained using rule 1 from Table 7.7 and CAMD. There are two distillation
regions in this figure. Figure 7.16 shows the product sequence using the middle
vessel column for this mixture.

7.2.2   Entrainer Selection Rules for Heterogeneous Batch Distillation

Heterogeneous batch distillation involves an entrainer that is partially miscible with one of the initial binary mixture components. Heterogeneous batch
distillation is advantageous as it provides more design alternatives for the
separation of an azeotropic binary mixture than with homogeneous batch distillation, presents simplified distillation sequences as a consequence of fewer
distillation tasks, and admits the possibility of the crossing of batch distillation boundaries due to the reflux of either one or both of the decanter phases.
However, although it presents a simplified distillation sequence, that does not
necessarily mean the lowest cost or the most efficient overall operation. Further
purification of the decanted phase rich in the component with finite immis-
cibility with the entrainer may be required. Rodriguez-Donis et al. (2001b)
presented rules for selecting entrainers for heterogeneous batch distillation
which are presented below.

The general rules to select a feasible heterogeneous entrainer for the separation of azeotropic and close-boiling binary mixtures are based on criteria
related to the entrainer boiling temperature and its capability to form binary or ternary azeotropes (heterogeneous or homogeneous) with each of the
original mixture components. The rules were established from the analysis
of all ternary residue curve diagrams relevant to heterogeneous systems. As
before, total reflux/reboil and infinite tray number are assumed. Rodriguez-
Donis et al. (2001b) established six general conditions for the selection of

TABLE 7.10

Selection of Entrainers Enabling the Separation of Minimum-Boiling-Temperature Azeotropic Binary Mixtures by Homogeneous Batch Distillation. Curved Boundaries Case (Rodriguez-Donis et al., 2001a)

	Entrainer		
	Low Boiling	Intermediate Boiling	High Boiling
1	1		
2	$3(A \leftrightarrow B)[sn]$		
3		$4[un]$	
4	$3(A) + 4[un]$		
5		$2(A \leftrightarrow B)[un]$	
6		$2(A[un \leftrightarrow s], B)[s \leftrightarrow un]$	
7		$2(A \leftrightarrow B)[un] + 3(B \leftrightarrow A)[sn]$	
8		$2(A \leftrightarrow B) + 4[un]$	
9	$3(A \leftrightarrow B) + 4[sn]$		
10		$2(B) + 4[s]$	$2(A \leftrightarrow B) + 4[s]$
11	$2(A, B)[un \leftrightarrow s] + 4[s^1 \leftrightarrow un]$	$2(A, B) + 4[s^2 \leftrightarrow un]$	$2(A, B) + 4[s^3 \leftrightarrow un]$
12	$2(A[s \leftrightarrow B[un \leftrightarrow s]) + 3(B[sn] \leftrightarrow A[s \leftrightarrow sn] + 4[un]$	$2(A \leftrightarrow B)[un] + 3(B \leftrightarrow A)[sn] + 4[un]$	$2(A \leftrightarrow B) + 3(B \leftrightarrow A)[sn] + 4[un]$
13		$2(A \leftrightarrow B)[un] + 3(B \leftrightarrow A) + 4[sn]$	

TABLE 7.11

Selection of Entrainers Enabling the Separation of Maximum-Boiling-Temperature Azeotropic Binary Mixtures by Homogeneous Batch Distillation. Curved Boundaries Case (Rodriguez-Donis et al., 2001a)

	Low Boiling	Entrainer Intermediate Boiling	High Boiling
1			1
2			$2(A \leftrightarrow B)[un]$
3			$2(A \leftrightarrow B) + 4[un]$
4		$4[sn]$	
5			$2(B) + 4[sn]$
6		$3(A \leftrightarrow B)[sn]$	
7		$3(A[sn \leftrightarrow s], B[s \leftrightarrow sn])$	
8		$2(A \leftrightarrow B)[un] + 3(B \leftrightarrow A)[sn]$	
9		$3(A \leftrightarrow B) + 4[sn]$	
10	$3(A \leftrightarrow B) + 4[s]$	$3(A) + 4[s]$	
11	$3(A, B[sn \leftrightarrow s]) + 4[s^1 \leftrightarrow sn]$	$3(A, B[sn \leftrightarrow s]) + 4[s^2 \leftrightarrow sn]$	$3(A, B[sn \leftrightarrow s]) + 4[s^3 \leftrightarrow sn]$
12		$2(A \leftrightarrow B) + 3(B \leftrightarrow A)[sn] + 4[un]$	
13	$2(A \leftrightarrow B)[un] + 3(B \leftrightarrow A) + 4[sn]$	$2(A[un] \leftrightarrow B[un \leftrightarrow s]) + 3(B[s] \leftrightarrow A[s \leftrightarrow sn]) + 4[sn]$	

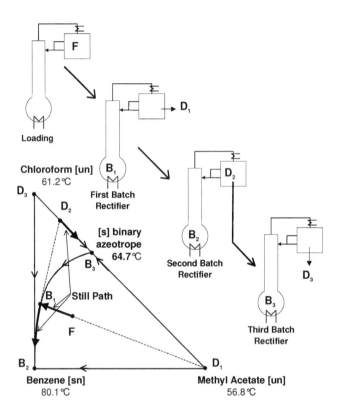

FIGURE 7.14
Example of Separation of a Maximum-Boiling Azeotrope Using a Heavy En-
trainer in a Homogeneous Batch Column with Curved Boundary (Rodriguez-
Donis et al., 2001a)

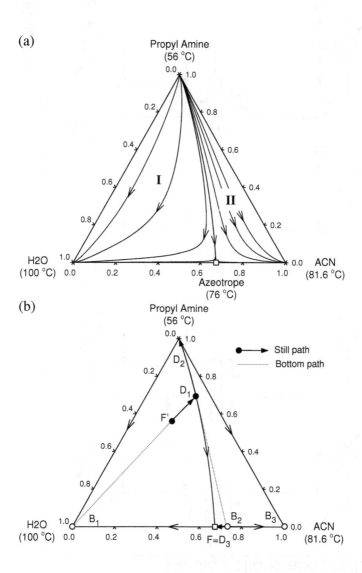

FIGURE 7.15
Residue Curve Maps (a) and Product Sequences (b) of the Acetonitrile–
Water–Propyl Amine System (Kim et al., 2004)

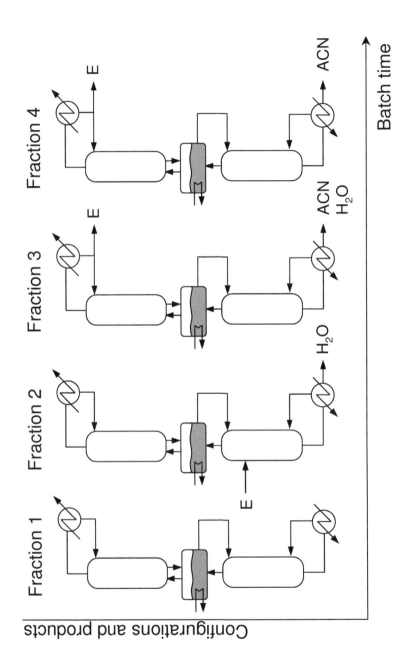

FIGURE 7.16
Operational Fractions of Batch Extractive Distillation in a Middle Vessel Column (Kim et al., 2004)

heterogeneous entrainers, allowing a feasible separation of minimum-boiling azeotropic binary mixtures, maximum-boiling azeotropic binary mixtures, and close-boiling binary mixtures. The first four conditions are identical to those defined for homogeneous batch distillation. These six general conditions are as follows:

1. No additional azeotrope is introduced.

2. A minimum temperature binary homoazeotrope exists between E and either A or B.

3. A maximum temperature binary homoazeotrope exists between E and either A or B.

4. E forms either a homogeneous or heterogeneous ternary azeotrope.

5. E is not miscible with one of the components of the original binary mixture.

6. E forms a minimum temperature binary heteroazeotrope with A or B.

Again, Equation 7.10 is used for points not defined by the rules in the six tables given below for minimum-boiling azeotropes, maximum-boiling azeotropes, and close-boiling components.

A feasible entrainer provides a ternary diagram where the two original components are or are not in the same batch distillation region and either the heteroazeotrope or one of the original components is a node. When the heteroazeotrope is an unstable node, a batch rectifier as the first batch task is the preferred choice because it leads to the lowest number of batch tasks needed to obtain a sequence of high-purity products. Conversely, a stripper (rectifier) can be used when the original component is a stable (unstable) node. The rules for entrainer selection are divided by these tasks.

Minimum-Boiling Azeotrope Separation

Table 7.12 shows rules for minimum-boiling azeotropes when the heteroazeotrope is an unstable node and the rectifier is the preferred configuration and Table 7.13 presents rules where the stripper is the preferred configuration.

Figure 7.17 shows a ternary system that exhibits a ternary unstable heteroazeotrope placed in the same distillation region as the original binary homoazeotrope. This rectification process involves separation of 1-propanol-water using benzene. Benzene obeys rule 8 for light entrainers in Table 7.12. For a rectifying operation the heterogeneous unstable node moves up the column after a total reflux step, the entrainer-rich phase is refluxed, the other phase is removed as a distillate cut, and the still path can reach the miscible original component vertex at the end of the process by control of both the initial amount of entrainer and the reflux of the entrainer-rich phase.

Figure 7.18a shows the sequence of the heterogeneous batch rectification displayed in Figure 7.17. The unstable ternary heteroazeotrope boils overhead.

TABLE 7.12

Selection of Entrainers Enabling the Separation of Minimum-Boiling-Temperature Azeotropic Binary Mixtures by a Heterogeneous Batch Rectifier(Rodriguez-Donis et al., 2001b)

| | Entrainer | | |
	Low Boiling	Intermediate Boiling	High Boiling
1		$6(A \leftrightarrow B)[un]$	
2		$2(A \leftrightarrow B) + 6(B \leftrightarrow A)[un]$	
3	$3(A[sn \leftrightarrow s] \leftrightarrow B[sn]) + 6(B \leftrightarrow A)[un]$		$3(A \leftrightarrow B)[sn] + 6(B \leftrightarrow A)[un]$
4			$4[un] + 5(A \leftrightarrow B)$
5		$4[un \leftrightarrow s] + 6(A \leftrightarrow B)$	
6		$2(A \leftrightarrow B) + 4[un] + 5(B \leftrightarrow A)$	
7	$3(A) + 4[un] + 5(B)$	$3(A[s] \leftrightarrow B[sn]) + 4[un] + 5(B \leftrightarrow A)$	$3(A \leftrightarrow B)[sn] + 4[un] + 5(B \leftrightarrow A)$
8		$2(A \leftrightarrow B) + 4[un] + 6(B \leftrightarrow A)$	
9	$2(A[un^2 \leftrightarrow s^3] \leftrightarrow B[un^2 \leftrightarrow s^1])$ $+4[s] + 6(B \leftrightarrow A)[un]$	$2(A[un^1 \leftrightarrow s^3] \leftrightarrow B[un^1 \leftrightarrow s^2])$ $+4[s] + 6(B \leftrightarrow A)[un]$	$2(A[un^3 \leftrightarrow s^1] \leftrightarrow B[un^3 \leftrightarrow s^2])$ $+4[s] + 6(B \leftrightarrow A)[un]$
10	$3(A)[s] + 4[un] + 6(B)[un \leftrightarrow s]$	$3(A \leftrightarrow B)[sn] + 4[un] + 6(B \leftrightarrow A)$	
11		$3(A \leftrightarrow B)[s \leftrightarrow sn] + 4[sn \leftrightarrow s] + 6(B \leftrightarrow A)[un]$	
hline 12		$3(A \leftrightarrow B)[s \leftrightarrow sn] + 4[sn \leftrightarrow s] + 6(B \leftrightarrow A)[un]$	

TABLE 7.13

Selection of Entrainers Enabling the Separation of Minimum-Boiling-Temperature Azeotropic Binary Mixtures by a Heterogeneous Batch Stripper (Rodriguez-Donis et al., 2001b)

	Low Boiling	Entrainer Intermediate Boiling	High Boiling
1	$1+5(A \leftrightarrow B)$	$1+5(A)$	
2		$6(A \leftrightarrow B)[s]$	
3	$4[s]+5(A \leftrightarrow B)$		
4		$2(A \leftrightarrow B)[un \leftrightarrow s]+5(B \leftrightarrow A)$	
5		$2(A \leftrightarrow B)[un \leftrightarrow s]+6(B \leftrightarrow A)[s]$	
6		$2(A \leftrightarrow B)+4[s]+5(B \leftrightarrow A)$	
7		$3(A \leftrightarrow B)+6(B \leftrightarrow A)[s]$	
8	$3(A[sn \leftrightarrow s] \leftrightarrow B[sn])+5(B \leftrightarrow A)$	$3(A)[s]+5(B)$	
9	$3(A \leftrightarrow B)+4[sn \leftrightarrow s]+5(B \leftrightarrow A)$	$3(A)+4[s]+5(B)$	
10	$2(A^1 \leftrightarrow B^3)+4[s]+6(B \leftrightarrow A)[s]$	$2(A^2 \leftrightarrow B^3)+4[s]+6(B^2 \leftrightarrow A^3)[s]$	$2(A^2 \leftrightarrow B^1)+4[s]+6(B^2 \leftrightarrow A^1)[s]$

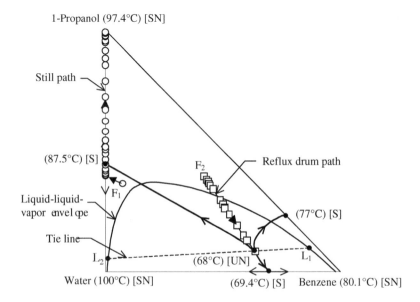

FIGURE 7.17
Separation of a Minimum-Boiling Azeotropic System by Heterogeneous Batch
Distillation (Reproduced Rodriguez-Donis et al., 2001b)

A liquid-phase split takes place into the decanter, and the water-rich phase L_2 is removed as the distillate P_1. The liquid-phase split depends on the 1-propanol mole fraction of the ternary heteroazeotrope. Although very little benzene is added to original binary mixture, the separation is possible due to the reflux of L_1, the entrainer-rich phase. The still path goes from the initial ternary composition F_1 toward the pure 1-propanol vertex (P_2). It crosses the distillation boundary between the 1-propanol-water binary homoazeotrope and the ternary heteroazeotrope. In this separation, a single batch rectification task allows the separation of pure 1-propanol, but several supplementary batch distillation tasks may be required if a water purity higher than that of L_2 is sought. Because water is a stable node, a single additional stripper will here allow one to obtain pure water from the L_2 mixture.

This separation can also be done in a stripper as 1-propanol is a stable node. The stripper tasks for this separation are shown in Figure 7.18b. Because the initial ternary mixture F_2 has to be located in the batch distillation region including the 1-propanol vertex, more entrainer is required than in the rectifier case. During stripping, 1-propanol is removed from the bottom as product P_1, while the reflux drum departs from F_2 and ends at the ternary heteroazeotrope. A liquid phase split takes place in the reflux drum. The process finishes when the reflux drum reaches the ternary heteroazeotrope. The aqueous phase L_2 is removed as product P_2, whereas the entrainer-rich phase L_1 can be recycled to the next batch task.

Maximum-Boiling Azeotrope Separation

Table 7.14 shows rules for maximum-boiling azeotropes when the heteroazeotrope is an unstable node and the rectifier is the preferred configuration and Table 7.15 presents rules where the stripper is the preferred configuration.

Separation of Close-Boiling Mixture

Close-boiling mixtures are as difficult to separate as azeotropic systems. The use of an entrainer that forms a new azeotrope that is easier to break than the original pinch binary is a typical solution to this problem. Table 7.16 shows rules for close-boiling azeotropes when the heteroazeotrope is an unstable node and the rectifier is the preferred configuration and Table 7.17 presents rules where the stripper is the preferred configuration.

Similar to the homogeneous entrainer used to separate the mixture shown in Figure 7.13, a heterogeneous entrainer, tolune, can also be used to separate the same mixture of water and acetic acid, as shown in Figure 7.19. This entrainer obeys rule 2 in Table 7.16 because toluene forms an unstable heteroazeotrope with water and a homogeneous minimum-boiling azeotrope with acetic acid. The rectification sequence is shown in Figure 7.18a. The initial ternary mixture F is homogeneous and contains a small amount of entrainer. After the infinite reflux operation, the entrainer stays mainly on the column trays and in the decanter. A single batch rectifier allows the separation of

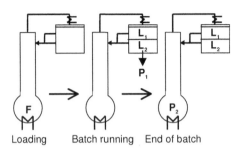

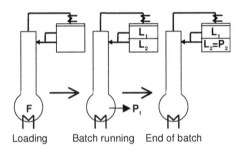

FIGURE 7.18

Separation Tasks for Heterogeneous Batch Distillation (Rodriguez-Donis et al., 2001b)

TABLE 7.14

Selection of Entrainers Enabling the Separation of Maximum-Boiling-Temperature Azeotropic Binary Mixtures by Heterogeneous Batch Rectifier (Rodriguez-Donis et al., 2001b)

	Entrainer		
	Low Boiling	Intermediate Boiling	High Boiling
1	$6(A \leftrightarrow B)$	$6(A)$	
2	$2(A \leftrightarrow B) + 6(B \leftrightarrow A)[un]$		$2(A \leftrightarrow B)[un \leftrightarrow s] + 6(B \leftrightarrow A)[un]$
3	$4[sn \leftrightarrow un] + 6(A \leftrightarrow B)[un \leftrightarrow s]$	$4[sn \leftrightarrow un] + 6(A)[un \leftrightarrow s]$	
4	$2(A \leftrightarrow B)[s] + 4[un] + 5(B \leftrightarrow A)$	$2(A)[s] + 4[un] + 5(B)$	$2(B) + 4[un] + 5(A)$
5	$2(A \leftrightarrow B)[s] + 4[un] + 6(B \leftrightarrow A)[s]$		$2(A \leftrightarrow B)[s \leftrightarrow un \leftrightarrow s] + 4[un] +$ $6(B \leftrightarrow A)[un \leftrightarrow un \leftrightarrow s]$
6	$2(A \leftrightarrow B)[s \leftrightarrow un] + 4[sn \leftrightarrow s] + 6(B \leftrightarrow A)[un]$		

TABLE 7.15

Selection of Entrainers Enabling the Separation of Maximum-Boiling-Temperature Azeotropic Binary Mixtures by a Heterogeneous Batch Stripper (Rodriguez-Donis et al., 2001b)

	Entrainer		
	Low Boiling	Intermediate Boiling	High Boiling
1			$1 + 5(A \leftrightarrow B)$
2		$6(B)[s]$	$6(A[un \leftrightarrow B[un \leftrightarrow s])$
3		$2(B)[s] + 5(A)$	$2(A[un] \leftrightarrow B[un \leftrightarrow s]) + 5(B \leftrightarrow A)$
4		$3(A \leftrightarrow B)[sn \leftrightarrow s] + 5(B \leftrightarrow A)$	
5		$3(A \leftrightarrow B)[sn \leftrightarrow s] + 6(B \leftrightarrow A)[un]$	
6		$3(A) + 6(B)[s]$	
7		$4[sn] + 5(A \leftrightarrow B)$	$4[s] + 5(A \leftrightarrow B)$
8		$4[sn \leftrightarrow s] + 6(B)[s \leftrightarrow un]$	$4[sn] + 6(B)$
9			$4[un] + 6(A)[s]$
10			$4[s] + 6(A \leftrightarrow B)[un]$
11		$2(B)[un \leftrightarrow s] + 4[s \leftrightarrow sn] + 5(A)$	$2(B) + 4[sn] + 5(A)$
12			$2(A \leftrightarrow B) + 4[s] + 5(B \leftrightarrow A)$
13		$3(A)[sn \leftrightarrow s] + 4[sn] + 6(B)[s]$	
14		$3(A \leftrightarrow B) + 4[s] + 5(B \leftrightarrow A)$	
15		$3(A \leftrightarrow B)[sn \leftrightarrow s] + 4[un] + 6(B \leftrightarrow A)[s]$	
16		$3(A \leftrightarrow B) + 4[s] + 6(B \leftrightarrow A)$	

TABLE 7.16
Selection of Entrainers Enabling the Separation of Close-Boiling Binary Mixtures by a Heterogeneous Batch Rectifier (Rodriguez-Donis et al., 2001b)

	Entrainer		
	Low Boiling	Intermediate Boiling	High Boiling
1	$6(A \leftrightarrow B)$	$6(A)[un]$	
2	$2(A \leftrightarrow B) + 6(B \leftrightarrow A)[un]$		
3	$3(B) + 6(A)$		
4	$4[un] + 6(A \leftrightarrow B)$	$4[un] + 6(A)$	
5	$2(A \leftrightarrow B) + 4[un] + 5(B \leftrightarrow A)$	$2(A) + 4[un] + 5(B)$	
6	$2(A \leftrightarrow B)[un \leftrightarrow s] + 4[s \leftrightarrow un] + 6(B \leftrightarrow A)[un \leftrightarrow s]$		
7	$3(B)[sn \leftrightarrow s] + 4[un \leftrightarrow sn] + 6(A)[s \leftrightarrow un]$		

water $(P_1 = L_2)$. Acetic acid is obtained in the still (P_2) at the process end. This mixture can also be separated with a stripper because the acetic acid vertex is a stable node.

7.2.3   Feasibility of Extractive Batch Distillation

The extractive distillation process using the rectifier consists of four steps which are: (1) operation under total reflux without solvent feeding, (2) operation under total reflux with solvent feeding, (3) operation under finite reflux with solvent feeding, and (4) operation under finite reflux without solvent feeding.

However, the operation steps and the size of the reboiler can be reduced if the middle vessel column is used. Safrit et al. (1995) and Safrit and Westerberg (1997) investigated batch extractive distillation in the middle vessel column. They showed that the extractive process is comprised of two steps (Operations 2 and 3) and requires a much smaller still pot size. They also identified feasible and infeasible regions, and showed that by varying column conditions such as the product rate, reflux ratio, and reboil ratio, one can 'steer' the middle vessel composition to escape from an infeasible region if it is located in one.

Lelkes et al. (1998) studied the feasibility of extractive batch distillation extensively. They stated that a theoretical recovery of 100 percent can only be possible when the steering is toward the simplex of the intermediate component on the triangular diagram. The authors found that the necessary and sufficient condition for feasibility is to have at least one connecting route between the still path and the rectifying profile in step 2. It is the extractive profile that connects the profiles (Figure 7.20). By varying the ratio of the entrainer flow rate to the vapor flow rate (E/V), one can change the extractive profile and thus find feasible operating conditions. They suggested that step 2 is important because the azeotrope is broken at this stage by the entrainer.

TABLE 7.17

Selection of Entrainers Enabling the Separation of Close-Boiling Binary Mixtures by a Heterogeneous Batch Stripper (Rodriguez-Donis et al., 2001b)

	Entrainer		
	Low Boiling	Intermediate Boiling	High Boiling
1		$6(B)[s]$	
2		$1 + 5(A \leftrightarrow B)$	
3	$2(A \leftrightarrow B) + 5(B \leftrightarrow A)$	$2(A[un] \leftrightarrow B[s]) + 5(B \leftrightarrow A)$	$2(B)[s] + 5(A)$
4		$2(A \leftrightarrow B)[un] + 6(B \leftrightarrow A)$	
5	$3(A)[s] + 5(B)$	$3(A[s] \leftrightarrow B[sn]) + 5(B \leftrightarrow A)$	$3(A \leftrightarrow B)[sn] + 5(B \leftrightarrow A)$
6	$3(A) + 6(B)[un]$	$3(A)[s \leftrightarrow sn] + 6(B)$	$3(A)[sn] + 6(B)$
7		$4[s] + 6(B)$	
8		$2(B) + 4[s] + 5(A)$	
9	$3(A) + 4[s] + 5(B)$	$3(A) + 4[s] + 5(B)$ $3(B) + 4[sn] + 5(A)$	$3(A \leftrightarrow B) + 4[sn] + 5(B \leftrightarrow A)$
10	$3(A) + 4[un \leftrightarrow s] + 6(B)$	$3(A) + 4[s \leftrightarrow un \leftrightarrow sn] + 6(B)$	$3(A) + 4[s \leftrightarrow sn] + 6(B)$

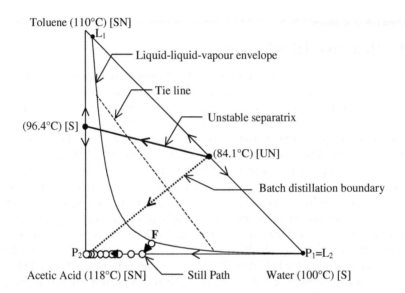

Toluene (110°C) [SN]

Liquid-liquid-vapour envelope

Tie line

(96.4°C) [S]

Unstable separatrix

(84.1°C) [UN]

Batch distillation boundary

F

P₂

Acetic Acid (118°C) [SN]   Still Path   Water (100°C) [S]

P₁=L₂

FIGURE 7.19
Separation of a Close-Boiling Mixture Using a Low Boiling Entrainer in a Heterogeneous Batch Column (Rodriguez-Donis et al., 2001b)

It should be remembered that since this feasibility analysis is based on the batch rectifier, not the middle vessel column, it also requires four steps and may require a large-sized reboiler.

Recently, Rodriguez-Donis et al. (2009a, 2009b) showed that the feasibility of extractive batch distillation (in terms of design and operating parameters) can be explained using a combination of the residue curve map, and the unidistribution and univolatility curves for heavy entrainers. Recently, their group (Barreto et al., 2011) also presented optimization of heterogeneous batch extractive distillation.

7.3   Reactive Distillation

Combining reaction and distillation is an old idea that has received renewed attention (Doherty and Malone, 2001). The improvements in some processes via reactive distillation are dramatic. However, combining reaction and distillation is not always advantageous. Conventional batch distillation with a chemical reaction (reaction and separation taking place in the same vessel and hence referred to as Batch REActive Distillation-BREAD) is particularly suitable when one of the reaction products has a lower boiling point than other products and reactants. The higher volatility of this reaction product increasing the liquid temperature and hence reaction rate, in the case of an irreversible reaction. With a reversible reaction, elimination of products by distillation favors the forward reaction. In both cases, a higher amount of distillate (proportional to the increase in conversion of the reactant) with the desired purity is expected than in distillation alone (Mujtaba, 2004). This section presents the design of batch reactive distillation using varying degrees of model complexity for the conventional batch column, followed by a section on batch distillation synthesis for reactive distillation for various configurations of batch columns.

7.3.1   Simplified Model

Recently, Huerta-Garrido et al. (2004) presented a simplified McCabe–Thiele method for conventional batch reactive distillation. This method provides insight into the operation of batch reactive distillation similar to the theoretical analysis presented in Chapter 2 and is presented below. This section and the examples are derived from Huerta-Garrido et al. (2004). For equilibrium reactions, the vapor-liquid equilibria can be combined with chemical equilibria to obtain the equilibrium relationship. Except for this relationship, the other equations for non-reactive batch distillation remain the same and hence all the methods presented earlier in this book are applicable to the equilibrium reactive batch distillation column. However, for non-equilibrium reac-

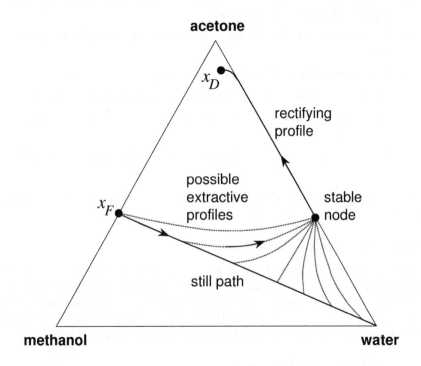

FIGURE 7.20
Possible Extractive Profiles for a Feasible Operation in Step 2

tive distillation, kinetics need to be considered and this changes the equations and methods. This section describes the McCabe-Theile method presented by Huerta-Garrido et al. (2004) for these systems. The method is presented for the constant reflux condition but can be extended to other operating conditions. Two cases are considered depending on whether the reaction takes place in the liquid phase or the vapor phase.

The following assumptions are used to derive the equations for kinetically controlled reactive batch columns:

1. Heat of mixing, stage heat losses, and sensible heat changes of both liquid and vapor are negligible.

2. The pressure on the column remains constant through the operation.

3. Startup conditions will not be considered.

4. The stages of the column will be considered in a quasi-steady state.

5. The reaction takes place on all of the stages but the reboiler and condenser.

6. The reaction is irreversible because of the continuous removal of the product.

If assumption 1 holds, then the molar flow rate of the phase in which a reaction does not occur can be considered as a constant.

7.3.1.1 Reaction in Liquid Phase

In Chapter 2, it was presented that for a binary system, if we separate the reboiler dynamics from the quasi-steady state assumption for the column, the functional relationship between distillate composition x_D and bottom composition x_B can be evaluated based on the McCabe–Theile method for continuous distillation. A similar relationship can be derived for reactive distillation.

Functional Relationship Between x_D and x_B

Consider the column shown in Figure 7.21. Since the reaction is occurring in the liquid phase, we can assume the vapor rate to be constant and equal to V, as shown in Figure 7.21.

The overall material balance around the condenser results in:

$$V = L_0 + D = D(R + 1) \tag{7.11}$$

In Figure 7.21 $\epsilon_{k,j}$ represents the sum of the molar turnover flow rate for component k (reactant defined as the base component) on stage j. The subtractions indicated in the figure are used to emphasize the contributions of reaction on each stage because $\epsilon_{k,j}$ is an accumulated sum.

For a binary system involving components k (heavy reactant) and i (light

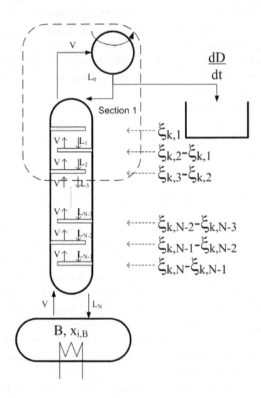

FIGURE 7.21
Schematic of Reactive Batch Distillation for a Liquid Phase Reaction

product), the overall material balance and the component balance around the top section and the j-th plate is given below.

$$y_{j+1}^{(i)} = \frac{L_j}{V}x_j^{(i)} + \frac{dD/dt}{V}x_D^{(i)} - \frac{\gamma_i}{-\gamma_k}\frac{\epsilon_{k,j-1}}{V} \qquad (7.12)$$

$$V = L_j + \frac{dD}{dt} - \gamma_T\epsilon_{k,j-1} \qquad (7.13)$$

where γ_i represents the stoichiometric coefficients of reaction (negative for reactant k and positive for product i), γ_T is the total sum of stoichiometric coefficients, and $(\gamma_i/-\gamma_k)\epsilon_{k,j}$ represents the production of i due to the reaction of k on stage j. The last of the terms in Equation 7.13 represents the overall change in the number of moles due to the reaction. Notice that, although we are using a binary system as the basis for our derivations here, the equations are written for multicomponent system so that they can be applied to a multicomponent systems.

For a system where there is no change in moles between reactants and products ($\gamma_i = -\gamma_k$) then γ_T will be zero, resulting in a constant molar flow assumption for the liquid phase also. Then Equation 7.12 reduces to the following equation for a binary system.

$$y_{j+1} = \frac{R}{R+1}x_j + \frac{1}{R+1}(x_D - \frac{\epsilon_{j-1}}{dD/dt}) \qquad (7.14)$$

If you compare Equation 2.7 in Chapter 2 with Equation 7.14, the slope of the line is the same as before except the intercept is changing for each plate. Consider the McCabe–Thiele diagram shown in Figure 7.22. Equation 7.14 represents several operating lines parallel to each other because, for the same slope, the y intercept is changing in terms of the accumulated sum of the reaction molar turnover flow rates. Starting from the top at the distillate composition, x_D, the first operating line intersects the 45 line reaction molar flow rate. Then, as we move down the column, the accumulated sum of the reaction molar turnover flow rates increases and the intersection point also moves further to the left. These intersection points are the reactive difference points and can be calculated from Equation 7.15. Thus, combining the reactive difference point with the operating line for non-reactive batch distillation, one can derive the operating lines for reactive distillation.

$$x_j = x_D - \frac{\epsilon_j}{dD/dt} \qquad (7.15)$$

For the case where there is a change in number of moles due to the reaction, the liquid flowrate on plate j changes as given below. This changes the reflux on each plate. Equation 7.12 is directly used for each plate along with the following equation for the McCabe–Thiele calculation.

$$L_j = L_{j-1} + \gamma_t(\epsilon_{k,j} - \epsilon_{k,j-1}) \qquad (7.16)$$

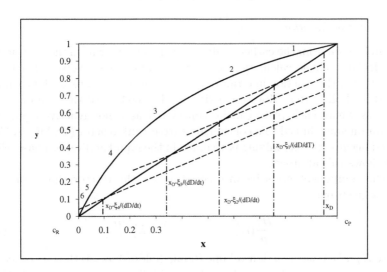

FIGURE 7.22
McCabe–Thiele Diagram for Reactive Batch Distillation for a Liquid Phase
Reaction

Reboiler Dynamics

To derive the reboiler dynamics, overall material balance and component balance around the column are used. These equations are given below.

$$\frac{dB}{dt} = -\frac{dD}{dt} + \gamma_T \epsilon_{k,T} \tag{7.17}$$

$$\frac{dBx_B^{(i)}}{dt} = -\frac{dD}{dt}x_D^{(i)} + \frac{\gamma_i}{-\gamma_k}\epsilon_{k,T} \tag{7.18}$$

where $\epsilon_{k,T}$ is the total molar turnover flow rate (k being the base component). Substituting $dD/dt = V/(R+1)$ results in the following equations.

$$\frac{dB}{dt} = -\frac{V}{R+1} + \gamma_T \epsilon_{k,T} \tag{7.19}$$

$$\frac{x_B^{(i)}}{dt} = \frac{1}{B}\left(\frac{V}{R+1}(x_B^{(i)} - x_D^{(i)}) + [\frac{\gamma_i}{-\gamma_k} - \gamma_T x_B^{(i)}]\epsilon_{k,T}\right) \tag{7.20}$$

Compared to a normal batch distillation column without a reaction, these equations have an extra term related to reactions.

Extent of the Reaction

As can be seen from the earlier analysis, the reaction terms in the equations for the McCabe–Thiele calculation and reboiler dynamics involve calculation of the extent of the reaction in terms of molar turnover flowrate $\epsilon_{k,j}$ for each plate and the reboiler. Here Huerta-Garrido et al. (2004) used the idea of reactive cascades where each plate can be considered as a non-reactive equilibrium separation stage linked to a continuous stirred tank reactor (CSTR). A CSTR is used because perfect mixing is assumed in the liquid and vapor phase. Figure 7.23 shows this approach.

The component mass balance equation for the reactor shown in Figure 7.23 is given by:

$$\frac{H_j}{\rho}(r_{k,j}) = L_j x_J^{(k)} - L_j' x_J^{'(k)} \tag{7.21}$$

where H_j/rho is the reaction volume (liquid holdup), $r_{k,j}$ is the rate of reaction of component k on stage j, L is the liquid flow rate, and x is the liquid composition.

The reactor outlet liquid stream, L_j', can be expressed in terms of the molar conversion fraction of the reactant k, C_j:

$$L_j' x_J^{'(k)} = L_{j-1} x_{j-1}^{(k)}(1 - C_j) \tag{7.22}$$

Substituting in Equation 7.21:

$$\frac{H_j}{\rho}(r_{k,j}) = L_{j-1}' x_{j-1}^{'(k)} C_j \tag{7.23}$$

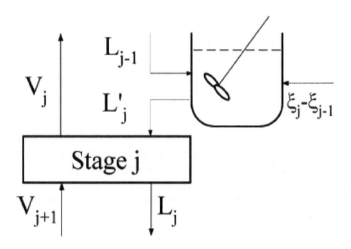

FIGURE 7.23
Reaction in Liquid Phase, Cascade Approach

Assuming an elemental irreversible reaction and limiting reactant k, the rate of reaction can written as:

$$r_{k,j}/\rho = K_j(\frac{L'_j x_j^{'(k)}}{Vol})^{\gamma_k} \tag{7.24}$$

where the term inside the parenthesis represents the concentration of the reactant k and K_j is a kinetic parameter given by the Arrehenius equation. Vol is the volumetric flow rate entering the stage and can be calculated from the liquid flowrate L_{j-1}.

$$Vol = L_{j-1}/\rho \tag{7.25}$$

where ρ is the density.

Combining equations 7.23-7.25, an implicit equation for molar conversion of the reactant can be obtained.

$$H_j K_j(\frac{L_{j-1}x_{j-1}^{(k)}(1-C_j)}{Vol})^{\gamma_k} = L'_j x_j^{'(k)}C_j \tag{7.26}$$

Once C_j can be calculated from Equation 7.26, the molar turnover flow rate can be calculated through the following expression.

$$\epsilon_{k,J} = \sum_{j=1}^{J} L'_{j-1}x_{j-1}^{'(k)}C_j \tag{7.27}$$

The McCabe–Thiele method along with the reboiler dynamics and extent of reaction calculations can be used iteratively for the constant reflux, variable reflux, and optimal reflux operation, as presented in Chapter 2.

Example 7.3: Consider the example corresponding to an isomerization reaction in the liquid phase:

$$o - xylene(C_8H_{10}) \Rightarrow ethylbenzene(C_8H_{10})$$

The lightest component is ethylbenzene. The data used for this example are shown in Table 7.18. Find the effect of reflux ratio on average distillate composition.

Solution: Figure 7.24 shows the McCabe–Thiele diagram at the first step. As expected, the contribution of the reaction is greater for stages closer to the reboiler (larger reactant concentrations). Figure 7.24 does not show the last of the operating lines because the difference point is negative. Figure 7.25 presents three instances of the calculation of the average distillate composition by changing the value of the reflux ratio. Because the reaction shows high conversion values, the effect of the reflux ratio is not significant. Observe also that, because of the reaction, the average distillate composition increases with time; this result is contrary to the behavior observed in nonreactive batch distillation columns.

Example 7.4: Consider the the reaction isobutyl-isobutyrate $(C_8H_{16}O_2)$ going to 2-isobutyraldehyde (C_4H_8O) in the liquid phase. In this reaction the number of

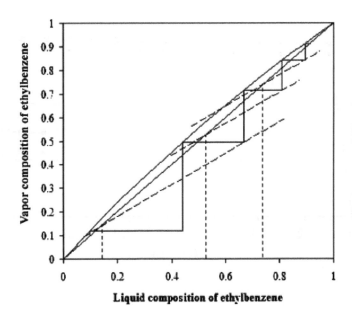

FIGURE 7.24
McCabe–Thiele Diagram for Example 7.3 (Huerta-Garrido et al., 2004)

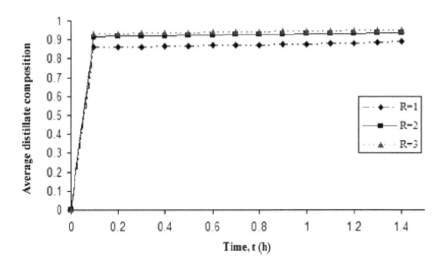

FIGURE 7.25
Average Distillate Composition Profile for Ethylbenzene (Huerta-Garrido et
al., 2004)

TABLE 7.18
Data for the Reaction-Separation System o-Xylene–Ethylbenzene

Parameter	Value
No. of Stages	4 +reboiler
Feed	100 $kmol$
Feed Comp. (ethylbenzene)	0.1
Vapor Boilup Rate	140 $kmolh^{-1}$
Column Pressure	1 atm
Batch Time	1.4 h
Order of Reaction	1
Activation Energy	30,000 $kjkmol^{-1}$
Frequency Factor	1.6 E9 h^{-1}
Volumetric Flow Rate	2 m^3h^{-1}
Liquid Holdup	4 E-4 m^3

moles changes. The lightest component is isobutyraldehyde. Kinetics and equilibrium data are shown in Table 7.19. Find the average distillate composition profile for various reflux ratios.

TABLE 7.19
Data for the Reaction-Separation System Isobutyl–Isobutyrate–Isobutyraldehyde

Parameter	Value
No. of Stages	3 +reboiler
Feed	100 $kmol$
Feed Comp. (Isobutyraldehyde)	0.1
Vapor Boilup Rate	140 $kmolh^{-1}$
Column Pressure	1 atm
Batch Time	1.4 h
Order of Reaction	1
Activation Energy	30,000 $kjkmol^{-1}$
Frequency Factor	1.0 E3 h^{-1}
Volumetric Flow Rate	2 m^3h^{-1}
Liquid Holdup	4 E-4 m^3

Solution: Figure 7.26 shows the McCabe–Thiele diagram for the operation time of 0.6 h. The distillate composition is very close to 1. Changes in the number of moles present an impact on the difference points as well as on the slope of the operating lines. Figure 7.27 shows the average distillate composition profiles for three values of the reflux ratio; because the liquid flow rates are changing from one stage to another, the reflux ratio has a significant effect in the composition. Also, in this example separation predominates over the reaction and the average distillate composition of isobutyraldehyde decreases with time.

7.3.1.2   Reaction in Vapor Phase

Again considering the quasi-steady state assumption for plates and the reboiler dynamics for a reaction in vapor phase, the following model equations result.

Functional Relationship Between x_D and x_B

Consider the column shown in Figure 7.22. Since the reaction is occurring in the vapor phase, we can assume the liquid rate to be constant and equal to L, as shown in Figure 7.28.

The operating line for the McCabe–Thiele method for the vapor phase reaction shown in Figure 7.28 is given below.

$$y_{j+1} = \frac{V_1}{V_{j+1}}\frac{R}{R+1}x_j + \frac{V_1}{V_{j+1}}\frac{1}{R+1}(x_D - \frac{\epsilon_{j-1}}{dD/dt}) \qquad (7.28)$$

where changes in vapor flow rate due to the reaction are given by:

$$V_{j+1} = V_j + \gamma_T(\epsilon_{k,j} - \epsilon_{k,j-1}) \qquad (7.29)$$

Reboiler Dynamics

The reboiler dynamics for this case can be written as:

$$\frac{dB}{dt} = -\frac{V_1}{R+1} + \gamma_T\epsilon_{k,T} \qquad (7.30)$$

$$\frac{x_B^{(i)}}{dt} = \frac{1}{B}(\frac{V_1}{R+1}(x_B^{(i)} - x_D^{(i)}) + [\frac{\gamma_i}{-\gamma_k} - \gamma_T x_B^{(i)}]\epsilon_{k,T}) \qquad (7.31)$$

Extent of the Reaction

Again the approach is based on the phenomenon of decomposition of the reaction and separation phenomena, as shown in Figure 7.29. Similar to the liquid phase reaction, two equations for molar conversion, C_j, and for molar turnover rate, $\epsilon_{k,j}$, can be derived as given below.

$$H_j K_j(\frac{V_j' y_{j-1}'^{(k)}(1-C_j)}{Vol})^{\gamma_k}(1-C_j) = V_j' y_j'^{(k)}C_j \qquad (7.32)$$

$$\epsilon_{k,J} = \sum_{j=1}^{J} V_j' y_j'^{(k)}\frac{C_j}{1-C_j} \qquad (7.33)$$

Example 7.5: Consider the isomerization reaction n-butane $(C_4H_{10}) \Rightarrow$ isobutane (C_4H_{10}) taking place in the vapor phase; isobutane is the lightest component. Table 7.20 shows the parameters for this example.

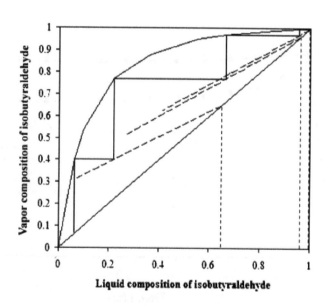

FIGURE 7.26

McCabe–Thiele Diagram for Example 7.4 (Huerta-Garrido et al., 2004)

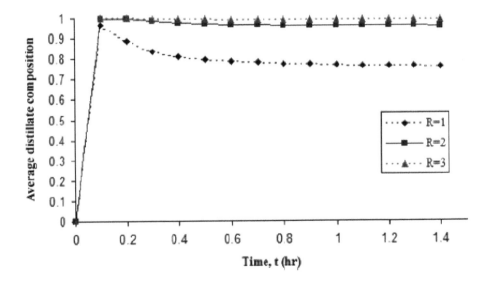

FIGURE 7.27
Average Distillate Composition Profile for Isobutyraladyde (Huerta-Garrido
et al., 2004)

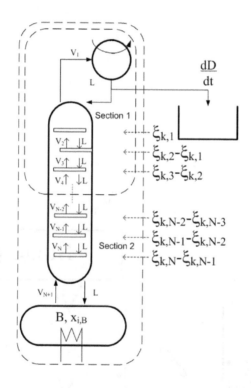

FIGURE 7.28
Schematic of Reactive Batch Distillation for a Vapor Phase Reaction

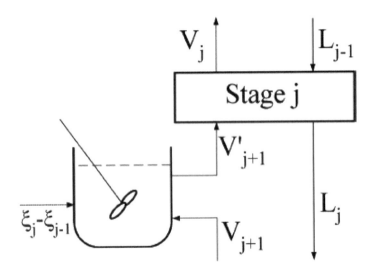

FIGURE 7.29
Reaction in Vapor Phase, Cascade Approach

TABLE 7.20

Data for the Reaction-Separation System Isobutyl–Isobutyrate–Isobutyraldehyde

Parameter	Value
No. of Stages	3 +reboiler
Feed	100 $kmol$
Feed Comp. (Isobutyraldehyde)	0.1
Vapor Boilup Rate	140 $kmol h^{-1}$
Column Pressure	1 atm
Batch Time	1.4 h
Order of Reaction	1
Activation Energy	30,000 $kjkmol^{-1}$
Frequency Factor	1.0 E3 h^{-1}
Volumetric Flow Rate	2 $m^3 h^{-1}$
Liquid Holdup	4 E-4 m^3

Solution: Figure 7.30 shows the McCabe–Thiele diagram for the operation time of 0.6 h. This problem has no change in moles due to the reaction so the operating lines are parallel to each. Once again, the last of the difference points is negative, and the last of the operating lines has not been drawn. The diagram corresponds to a reflux ratio equal to 2. Figure 7.31 shows the behavior of the average distillate composition of isobutane; the trend is similar to Example 7.3 as the average distillate composition increases with time because of the reaction and the reflux ratio has very little effect on the average product composition.

7.3.2 Rigorous Model

Similar to the rigorous model presented in Chapter 3, the rigorous model for reactive batch distillation is presented below and in Table 7.21. The first rigorous modeling study of reactive batch distillation was proposed by Cullie and Reklaitis (1986). They stated that for reactive batch distillation the finite difference assumption for heat balance can create problems if the heat changes due reactions are severe. In their study, they used the differential equation for heat balance and used a supercomputer to solve the problem. They did not use the constant molar holdup assumption. However, many later studies like Mujtaba and Macchietto (1997) use finite difference approximations for enthalpy balance and use constant molar holdup approximation for reactive batch distillation. We present both approaches here.

For the condenser:

$$L_0 = R\frac{dD}{dt} \tag{7.34}$$

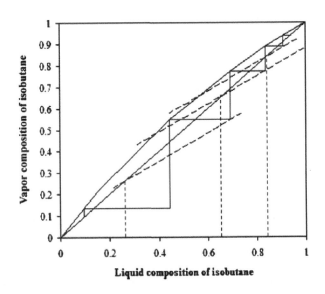

FIGURE 7.30
McCabe–Thiele Diagram for Example 7.5 (Huerta-Garrido et al., 2004)

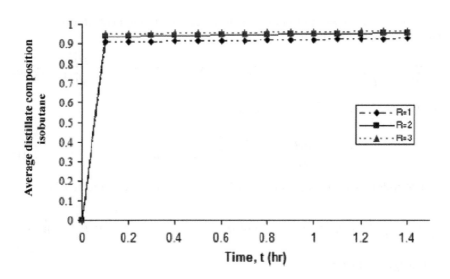

FIGURE 7.31
Average Distillate Composition Profile for Isobutane (Huerta-Garrido et al., 2004)

$$\frac{dH_D}{dt} = V_1 - \frac{dD}{dt}(R+1) + \triangle n_D H_D \tag{7.35}$$

$$\frac{dx_D^{(i)}}{dt} = \frac{V_1}{H_D}(y_1^{(i)} - x_D^{(i)}) + \frac{\gamma_i}{-\gamma_k}r_{k,D}/\rho - \triangle n_D x_D^{(i)} \tag{7.36}$$

$$\frac{dI_D}{dt} = \frac{1}{H_D}[V_1 J_1 - (R+1)\frac{dD}{dt}I_0 + Q_D - I_j\frac{dH_j}{dt}] \tag{7.37}$$

Overall material balance around plate j:

$$\frac{dH_j}{dt} = V_{j+1} + L_{j-1} - V_j - L_j + \triangle n_j H_j \tag{7.38}$$

Component balance for component i around plate j:

$$\frac{dx_j^{(i)}}{dt} = \frac{1}{H_j}[V_{j+1}(y_{j+1}^{(i)} - x_j^{(i)}) + L_{j-1}(x_{j-1}^{(i)} - x_j^{(i)}) - V_j(y_j^{(i)} - x_j^{(i)})]$$

$$+ \frac{\gamma_i}{-\gamma_k}r_{k,j}/\rho - \triangle n_j x_j^{(i)} \tag{7.39}$$

Energy balance around plate j (here we are including heat of formation in the component enthalpy):

$$\frac{dI_j}{dt} = \frac{1}{H_j}[V_{j+1}J_{j+1} + L_{j-1}I_{j-1} - V_j J_j - L_j I_j - I_j\frac{dH_j}{dt}] \tag{7.40}$$

where $j = 0, 1, \ldots, N$ and $i = 1, 2, \ldots, n$.
for the reboiler:

$$\frac{dB}{dt} = L_N - V_B + B\triangle n_B \tag{7.41}$$

The differential balance for the still composition of component i is

$$\frac{dx_B^{(i)}}{dt} = \frac{1}{B}[L_N(x_N^{(i)} - x_B^{(i)}) - V_B(y_B^{(i)} - x_B^{(i)})] + \frac{\gamma_i}{-\gamma_k}r_{k,B}/\rho - x_B^{(i)}\triangle n_B \tag{7.42}$$

Energy balance :

$$\frac{dI_B}{dt} = \frac{1}{B}[L_N I_N - V_B J_B - I_j\frac{dB}{dt} + Q_B] \tag{7.43}$$

where,

- $\triangle n_B$, total change in moles due to the reaction in the reboiler $= \sum_i^n \frac{\gamma_i}{-\gamma_k}r_{k,B}/\rho$

- $\triangle n_D$, total change in moles due to the reaction in the condenser $= \sum_i^n \frac{\gamma_i}{-\gamma_k}r_{k,D}$

- $\triangle n_j$, total change in moles due to the reaction on plate $j = \sum_i^n \frac{\gamma_i}{-\gamma_k} r_{k,j}$

- $r_{k,j}$ is the rate of the reaction of component k on stage j

- γ_i represents the stoichiometric coefficients of the reaction (negative for reactant and positive for product)

- γ_T is the total sum of stoichiometric coefficients

- ρ is the molar density

Equilibrium relations:

$$y_j^{(i)} = f((x_j^{(k)}, k = 1, \ldots, n), TE_j, P_j) \tag{7.44}$$

Enthalpy calculations:

$$I_j = f((x_j^{(k)}, j = 1, \ldots, n), TE_j, P_j) \tag{7.45}$$

$$J_j = f((y_j^{(k)}, j = 1, \ldots, n), TE_j, P_j) \tag{7.46}$$

Cullie and Reklaitis (1986) did not use the finite difference approximation for heat balance and they did not assume constant molar holdup. Therefore, in their calculation the molar enthalpy of liquid phase calculated by Equation 7.45 should be equal to the enthalpy calculated by integration. In their approach, the molar flowrates are implicitly calculated. When these assumptions are relaxed, the model used by Mujtaba and Macchietto (1997) results. This model is presented in Table 7.21.

Cullie and Reklaitis (1986) assumed the initial total reflux condition without any reaction. However, this assumption may not be valid and the reaction will take place at the total reflux condition. Therefore, Table 7.22 presents the dynamics for the initial total reflux condition. It is assumed that the still and column along with the condenser are charged with feed at its bubble point.

7.3.3   Synthesis

As stated earlier, if one of the reaction products is lower boiling than the reactants then the conventional batch column can be used. However, depending on the reaction, this may be true or not true. Therefore, it is very important to select the proper column for batch reactive distillation. Conventional batch distillation is not suitable when all reaction products have higher boiling temperature than those of the reactants. An inverted batch column is suitable for such a situation. For cases where some of the reaction products have higher and some lower boiling points than those of the reactants, then neither conventional nor the inverted batch column is suitable. For such reaction schemes, the middle vessel column (MVC) column will be the most suitable one because the light and heavy products can now be withdrawn simultaneously from the column, thus pushing the reaction further to the product side (Mujtaba, 2004).

TABLE 7.21
Complete Column Dynamics for BREAD

Assumptions
• Negligible vapor holdup,
• Adiabatic operation,
• Theoretical plates,
• Constant molar holdup,
• Finite difference approximations for the enthalpy changes.

Composition Calculations

Condenser and Accumulator Dynamics

$$\frac{dx_D^{(i)}}{dt} = \frac{V_1}{H_D}(y_1^{(i)} - x_D^{(i)}) + \frac{\gamma_i}{-\gamma_k}r_{k,D} - \triangle n_D x_D^{(i)}, i = 1, 2, \ldots, n$$

Plate Dynamics

$$\frac{dx_j^{(i)}}{dt} = \frac{1}{H_j}[V_{j+1}y_{j+1}^{(i)} + L_{j-1}x_{j-1}^{(i)} - V_j y_j^{(i)} - L_j x_j^{(i)}] + \frac{\gamma_i}{-\gamma_k}r_{k,j} - \triangle n_j x_j^{(i)},$$
$$i = 1, 2, \ldots, n; j = 1, 2, \ldots, N$$

Reboiler Dynamics

$$\frac{dx_B^{(i)}}{dt} = \frac{1}{B}[L_N(x_N^{(i)} - x_B^{(i)}) - V_B(y_B^{(i)} - x_B^{(i)})] + \frac{\gamma_i}{-\gamma_k}r_{k,B} - x_B^{(i)}\triangle n_B, i = 1, 2,$$

Flowrate Calculations

At the Top of the Column

$$L_0 = R\frac{dD}{dt}; V_1 = (R+1)\frac{dD}{dt} - H_D\triangle n_D$$

On the Plates (liquid phase reaction

$$L_j = V_{j+1} + L_{j-1} - V_j + H_j\triangle n_j; j = 1, 2, \ldots, N$$
$$V_{j+1} = \frac{1}{J_{j+1}-I_j}[V_j(J_j - I_j) + L_{j-1}(I_j - I_{j-1}) + H_j(\delta I_j + \triangle n_j I_j)]$$
$$j = 1, 2, \ldots, N$$

On the Plates (vapor phase reaction)

$$L_j = V_{j+1} + L_{j-1} - V_j; j = 1, 2, \ldots, N$$
$$V_{j+1} = \frac{1}{J_{j+1}-I_j}[V_j(J_j - I_j) + L_{j-1}(I_j - I_{j-1}) + H_j\delta I_j]$$
$$j = 1, 2, \ldots, N$$

At the Bottom of the Column

$$\frac{dB}{dt} = L_N - V_B + B\triangle n_B$$

Heat Duty Calculations

Condenser Duty

$$Q_D = V_1 J_1 - (R+1)\frac{dD}{dt}I_D - H_D\delta_t I_D$$

Reboiler Duty

$$Q_B = V_B(J_B - I_B) - L_N(I_N - I_B) + B\delta_t I_B$$

Thermodynamics Models

Equilibrium Relations

$$y_j^{(i)} = f((x_j^{(k)}, k = 1, \ldots, n), TE_j, P_j)$$

Enthalpy Calculations

$$I_j = f((x_j^{(k)}, j = 1, \ldots, n), TE_j, P_j)$$
$$J_j = f((y_j^{(k)}, j = 1, \ldots, n), TE_j, P_j)$$

TABLE 7.22

Initial Dynamics for BREAD

<table>
<tr><td align="center">

Composition Calculations

Condenser and Accumulator Dynamics

$$\frac{dx_D^{(i)}}{dt} = \frac{V_1}{H_D}(y_1^{(i)} - x_D^{(i)}) + \frac{\gamma_i}{-\gamma_k}r_{k,D} - \triangle n_D x_D^{(i)}, i = 1, 2, \ldots, n$$

Plate Dynamics

$$\frac{dx_j^{(i)}}{dt} = \frac{1}{H_j}[V_{j+1}y_{j+1}^{(i)} + L_{j-1}x_{j-1}^{(i)} - V_j y_j^{(i)} - L_j x_j^{(i)}] + \frac{\gamma_i}{-\gamma_k}r_{k,j} - \triangle n_j x_j^{(i)},$$

$$i = 1, 2, \ldots, n; j = 1, 2, \ldots, N$$

Reboiler Dynamics

$$\frac{dx_B^{(i)}}{dt} = \frac{1}{B}[L_N(x_N^{(i)} - x_B^{(i)}) - V_B(y_B^{(i)} - x_B^{(i)})] + \frac{\gamma_i}{-\gamma_k}r_{k,B} - x_B^{(i)}\triangle n_B, i = 1, 2, \ldots, n$$

</td></tr>
<tr><td align="center">

Flowrate Calculations

At the Top of the Column

$$L_0 = R\frac{dD}{dt}; V_1 = L_0 - H_D \triangle n_D$$

On the Plates (liquid phase reaction

$$L_j = V_{j+1} + H_j\triangle n_j; j = 1, 2, \ldots, N$$

$$V_{j+1} = \frac{1}{J_{j+1}-I_j}[V_j(J_j - I_j) + L_{j-1}(I_j - I_{j-1}) + H_j(\delta I_j + \triangle n_j I_j)]$$

$$j = 1, 2, \ldots, N$$

On the Plates (vapor phase reaction)

$$L_j = V_{j+1} + L_{j-1} - V_j; j = 1, 2, \ldots, N$$

$$V_{j+1} = \frac{1}{J_{j+1}-I_j}[V_j(J_j - I_j) + L_{j-1}(I_j - I_{j-1}) + H_j\delta I_j]$$

$$j = 1, 2, \ldots, N$$

At the Bottom of the Column

$$\frac{dB}{dt} = L_N - V_B + B\triangle n_B$$

</td></tr>
<tr><td align="center">

Heat Duty Calculations

Condenser Duty

$$Q_D = V_1 J_1 - (R+1)\frac{dD}{dt}I_D - H_D\delta_t I_D$$

Reboiler Duty

$$Q_B = V_B(J_B - I_B) - L_N(I_N - I_B) + B\delta_t I_B$$

</td></tr>
<tr><td align="center">

Thermodynamics Models

Equilibrium Relations

$$y_j^{(i)} = f((x_j^{(k)}, k = 1, \ldots, n), TE_j, P_j)$$

Enthalpy Calculations

$$I_j = f((x_j^{(k)}, j = 1, \ldots, n), TE_j, P_j)$$

$$J_j = f((y_j^{(k)}, j = 1, \ldots, n), TE_j, P_j)$$

</td></tr>
</table>

Similar to the rules presented for selection of the column for nonazeotropic reactive systems presented in the above paragraph, rules for an azeotropic batch reactive distillation system can be derived. Recently, Guo et al. (2003) presented feasibility studies for reactive azeotropic batch distillation systems and most of the material presented below is derived from this paper.

Binary Systems

In order to understand the feasibility of azeotropic batch reactive distillation processes, Guo et al. (2003) presented first the binary isomerization reaction of $R \Rightarrow P$. Figure 7.32 presents a binary system with a maximum-boiling azeotrope. In this system, product P is more volatile than reactant R. This figure illustrates how a batch rectifier avoids the maximum-boiling azeotrope. As the reaction occurs in the reboiler, the reboiler composition crosses the azeotropic composition and the rectifier produces pure product P at the top. Similarly, Figure 7.33 shows a batch stripper breaking a binary minimum-boiling azeotrope. In this system, product P is the heavy component, and reactant R is the light component. It can be seen that the overhead still composition crosses the azeotropic composition by reaction. Thus pure product P can be produced at the bottom of the column.

Now we reverse the situation and use the batch rectifier to take care of the minimum-boiling azeotrope by the reaction. Figure 7.34 shows how the minimum-boiling azeotrope could not be avoided by the batch rectifier. Similarly, the product from the batch stripper is the maximum-boiling azeotrope, as shown in Figure 7.35.

In summary, the batch rectifier can be used to avoid a maximum-boiling azeotrope and the batch stripper for a minimum-boiling azeotrope. However, it should be remembered that in order to break the azeotrope and produce the desired product, the azeotropic composition must lie within the forward reaction zone. As the middle vessel column has both sections, it can be used to avoid a maximum-boiling azeotrope in the rectifier section and a minimum-boiling azeotrope in the stripper section. For ternary and multicomponent system residue curve maps can be used for synthesis as given below.

Residue Curve Maps for the Reactive Batch Rectifier

As shown earlier, residue curve maps are derived from the simple distillation system which is equivalent to a single stage in a column. The equations for BREAD simple distillation can be derived from Equations 7.20 and 7.31 for reboiler dynamics. This is similar to Equation 7.2 for non-reactive batch distillation except we have to add the reaction terms. Here we have assumed a general reaction where the liquid or vapor moles will change with the reaction. The rate of change of moles i in liquid and the amount of liquid in the reboiler will change as follows.

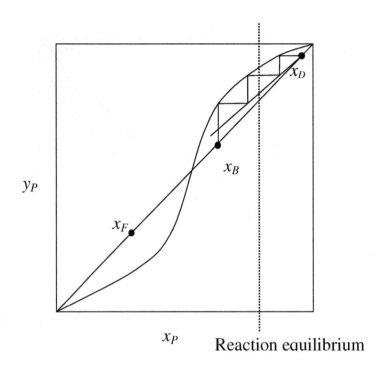

FIGURE 7.32
Batch Rectifier Analysis for a Maximum-Boiling Azeotropic System Using
McCabe–Thiele Diagrams (Guo et al., 2003)

$$\frac{dB}{dt} \quad = \quad -V + \gamma_T r_{k,1} B \qquad (7.47)$$

$$\frac{dBx^{(i)}}{dt} \quad = \quad -Vy^{(i)} + \gamma_i r_{k,1} B \qquad (7.48)$$

Combining the above two equations and putting warped time ζ using $d\zeta = dlnB$ results in the following equation.

$$\frac{dx^{(i)}}{d\zeta} \quad = \quad x^{(i)} - y^{(i)} + Da(\gamma_i - \gamma_T x^{(i)})RR \qquad (7.49)$$

where RR is the dimensionless rate of reaction $RR = r_{k,1}/r_0$, scaled by a

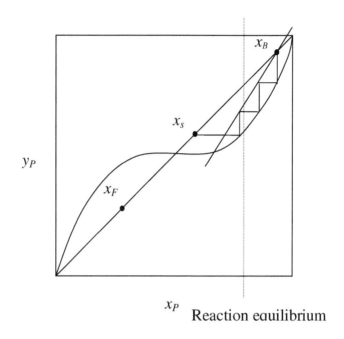

Reaction equilibrium

FIGURE 7.33
Batch Stripper Analysis for a Minimum-Boiling Azeotropic System Using
McCabe–Thiele Diagrams (Guo et al., 2003)

reference rate r_0. The dimensionless Da, the Damköhler number, gives the
ratio of the characteristic process time to the characteristic reaction time. For
a detailed discussion of this number please see Venimadhavan et al. (1994)
and Doherty and Malone (2001).

The choice of reference quantities for scaling is arbitrary, but some choices
are more useful than others when approximations of limiting cases are desired.
Sensible choices for r_0 can be the rate itself or the rate constant $r_0 = k_0$, evaluated at some standard temperature such as the lowest boiling point among
the pure component and azeotropes. When the reaction is catalyzed, r_0 typically also includes a reference catalyst concentration. It is desirable, but not
always possible, that a choice of a scaling factor for a dimensionless group
be made so that the magnitude of the terms in a model is reflected by the

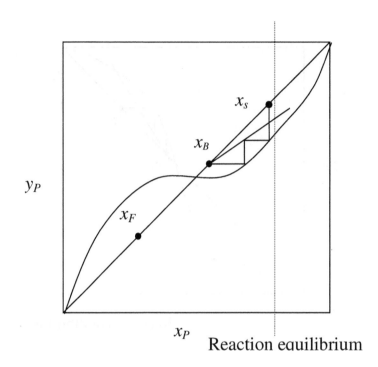

FIGURE 7.34
Batch Rectifier Analysis for a Minimum-Boiling Azeotropic System Using
McCabe–Thiele Diagrams (Guo et al., 2003)

numerical value of the group.

$$Da = \frac{t_{process}}{t_{reaction}} = \frac{B/V}{1/r_o} \tag{7.50}$$

Doherty and Malone (2001) discuss the importance of the Damköhler number, the discussion is given below.

For the ternary system, the effect of Da on residue curve maps is as follows. For small Da, the residue curves are identical to those for simple distillation without a reaction. For large Da, the residue curves are essentially straight lines for an initial point and rapidly approach the chemical equilibrium curve. At longer times, the rate decreases as the liquid composition approaches the reaction equilibrium curve, which is defined by $RR = 0$. When RR diminishes

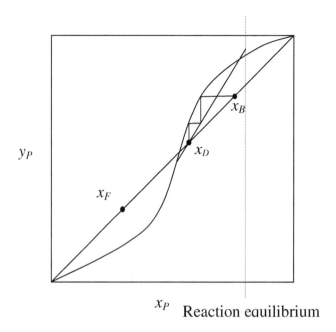

FIGURE 7.35
Batch Stripper Analysis for a Maximum-Boiling Azeotropic System Using
McCabe–Thiele Diagrams (Guo et al., 2003)

the first term in the residue curve map equation cannot be ignored. The composition change along the equilibrium curve is driven by both the first term (vaporization) and the reaction. Vaporization has a slow time scale ($t_{process}$), slower than the reaction time scale ($t_{reaction}$). In this case, the reaction adjusts quickly to changes in the liquid composition caused by vaporization. As the residue curve approaches singular points on the equilibrium curve, the composition reaches pure component or reactive azeotrope.

Guo et al. (2003) presented synthesis rules for various column configurations based on residue curve maps.

For the batch rectifier they used a simple reaction, $2B \rightleftarrows A + C$, for the study. In this reaction, A is assumed to be the light boiler, C is the heavy boiler, and reactant B is the intermediate reboiler. It is assumed that the

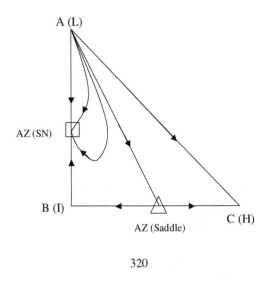

320

FIGURE 7.36
Residue Curve Map for a Batch Rectifier (Guo et al., 2003)

reaction occurs only in the reboiler for simplicity and quasi-steady state is
assumed for the column other than the reboiler. Figure 7.36 shows the residue
curve map (RCM) where there are two binary azeotropes: one is a maximum
boiling azeotrope (stable node) between A and B, while the other is a minimum boiling azeotrope (saddle) between B and C. Based on analysis of this
RCM, Guo et al. (2003) presented the first feasibility criteria for the batch reactive rectifier: when the light product is an unstable node and can be reached
from all distillation regions, a batch rectifier with a reaction can produce pure
products regardless of the number of azeotropes involved. In Figure 7.36 shows
two distillation regions. The initial feed charge in the still pot of the batch
rectifier is pure reactant B. Figure 7.37 shows that the reboiler composition
moves toward the reaction equilibrium curve as the reaction proceeds. Then it
approaches the vertex C, while product A is retrieved at the top and reactant
B is consumed. A is always the pure product regardless of the feed composition. B and C remain in the still and A is removed continuously from the
top. The reboiler composition moves toward the binary edge $B - C$. As B is
consumed, the still composition crosses the boundary and approaches vertex
C.

Figure 7.38 shows various RCMs where there are 1 to 4 azeotropes in a

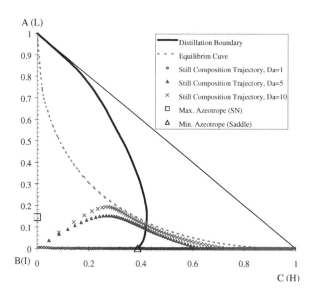

FIGURE 7.37
Simulation Results of the RCM Shown in Figure 7.36 (Guo et al., 2003)

system or binary and ternary azeotropes that follow the first feasibility criteria given below. For these cases the batch rectifier produces pure components.

If one of the reaction products is an unstable node but is not reachable by nonreactive distillation from all distillation regions, then a batch rectifier can still be used to obtain pure products. However, for that to be valid, the following two conditions are to be satisfied: (1) part of the reaction equilibrium curve lies in the same distillation region that the product does, and (2) Da is high enough to cross the distillation boundary. The RCM 420 shown in Figure 7.38 shows such a RCM. In this RCM, there are two maximum-boiling azeotropes: one is a stable node and the other is a saddle. These two azeotropes divide the RCM into two distillation region, as shown in Figure 7.39. The light product A is an unstable node and is not reachable from region I. It can be seen from Figure 7.39 that at high Da the still composition crosses the distillation boundary, enters region II, and then subsequently reaches vertex C. Figure 7.40 shows RCMs that satisfy these two feasiblity criteria to get pure components from the batch rectifier.

Residue Curve Maps for the Reactive Batch Stripper

The simple distillation analysis for the batch stripper with reactions results in following equations for RCM analysis.

$$\frac{dH_s}{dt} = -\frac{dB}{dt} + \gamma_T r_{k,1} H_s \tag{7.51}$$

$$\frac{dH_s x_s^{(i)}}{dt} = -\frac{dB}{dt} x_B^{(i)} + \gamma_i r_{k,1} H_s \tag{7.52}$$

Combining the above two equations and putting warped time ζ using $d\zeta = \frac{dB/dt}{H_s}$ results in the following equation.

$$\frac{dx_s^{(i)}}{d\zeta} = x_s^{(i)} - x_B^{(i)} + Da(\gamma_i - \gamma_T x_s^{(i)})RR \tag{7.53}$$

where H_s is the holdup in the still and x_s is the composition of the still.

Similar to the rectifier, Guo et al. (2003) derived feasibility criteria for the batch stripper to provide pure products based on the RCM analysis. The first feasibility criterion for the batch reactive stripper is that if one of the products is a stable node and can be reached from all distillation regions then the batch stripper can produce pure products. This is regardless of the number of azeotropes, types of azeotropes, or locations of azeotropes. Some of the RCMs that satisfy the first feasibility criterion when using the batch reactive stripper shown in Figure 7.41.

The second feasibility criterion deals with the condition if one of the product is a stable node but is not reachable from the distillation region to get pure products from this stripper, the RCM has to satisfy two conditions: (1) the reaction equilibrium curve has to lie in the same region as the stable node

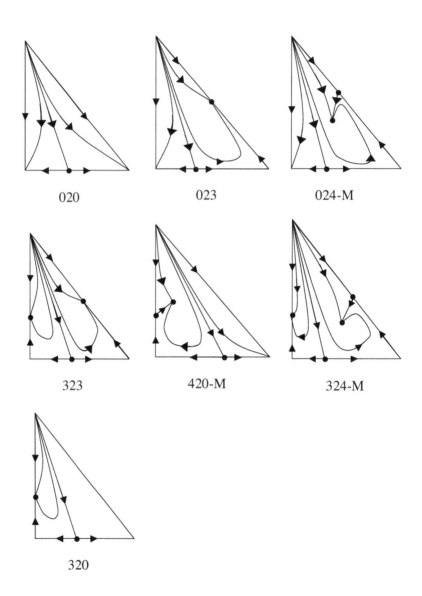

FIGURE 7.38
RCMs That Satisfy the First Feasibility Criterion for the Batch Rectifier (Guo et al., 2003)

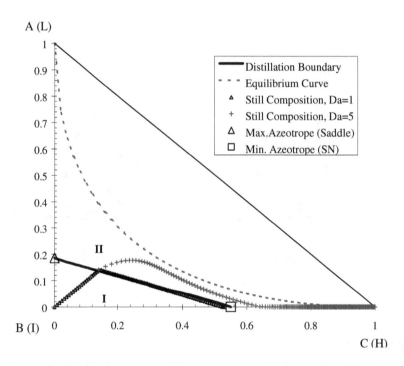

FIGURE 7.39
Simulation Results of RCM430 Shown in Figure 7.38 (Guo et al., 2003)

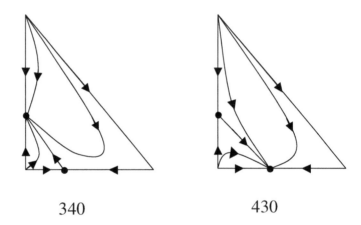

340   430

FIGURE 7.40
RCMs That Satisfy the Second Feasibility Criterion for the Batch Rectifier
(Guo et al., 2003)

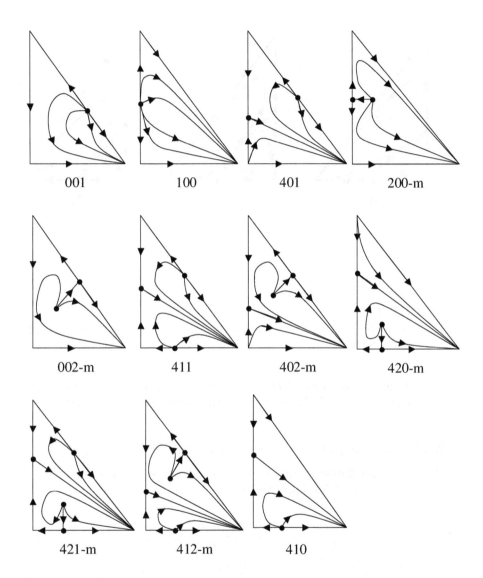

FIGURE 7.41
RCMs That Satisfy the First Feasibility Criterion for the Batch Stripper (Guo et al., 2003)

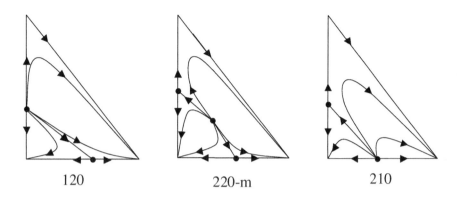

120   220-m   210

FIGURE 7.42
RCMs That Satisfy the Second Feasibility Criterion for the Batch Stripper
(Guo et al., 2003)

product and (2) the value of Da should be large enough to cross the distillation boundary. Some RCMs that satisfy the second feasibility criterion shown in Figure 7.42

Residue Curve Maps for the Reactive Batch Middle Vessel Column

The simple distillation system with the middle vessel with reactions can be represented by the following equations.

$$\frac{dH_m}{dt} = -\frac{dB}{dt} - \frac{dD}{dt} + \gamma_T r_{k,1} H_m \qquad (7.54)$$

$$\frac{dH_m x_m^{(i)}}{dt} = -\frac{dB}{dt} x_B^{(i)} - \frac{dD}{dt} x_D^{(i)} + \gamma_i r_{k,1} H_m \qquad (7.55)$$

Combining the above two equations and putting warped time ζ using $d\zeta = \frac{dB/dt + dD/dt}{H_s}$ results in the following equation.

$$\frac{dx_m^{(i)}}{d\zeta} = x_s^{(i)} - \frac{x_B^{(i)} + x_D^{(i)}}{2} + Da(\gamma_i - \gamma_T x_m^{(i)})RR \qquad (7.56)$$

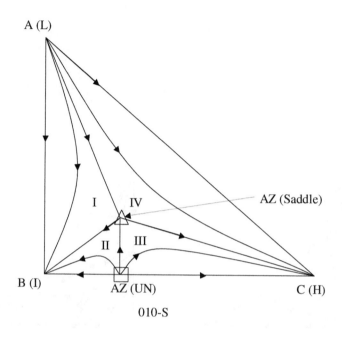

FIGURE 7.43
RCM for a Middle Vessel Column with a Reaction (Guo et al., 2003)

For the middle vessel with a reaction, Guo et al. (2003) presented the feasibility criteria for a system where reaction products are unstable and stable nodes lying in the same distillation region but are not reachable from the initial feed region, as shown in the RCM shown in Figure 7.43. In this case, the middle vessel column can produce pure products and is not restricted by the Da value. That is because in the middle vessel, the middle vessel holdup can be manipulated to cross the distillation boundary. Thus, the dynamics of the multivessel column provide more flexibility. Examples of RCMs for the middle vessel column where pure products can be produced are shown in Figure 7.44.

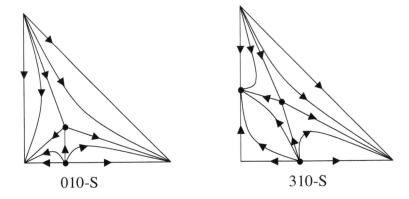

FIGURE 7.44
RCMs That Satisfy the Feasibility Criterion for the Middle Vessel Columnn
(Guo et al., 2003)

References

Ahmad, B. S. and P. I. Barton (1996), Homogeneous multicomponent azeotropic batch distillation, *AIChE J.*, **42** 3419.

Ahmad, B. S., Y. Zhang, and P. I. Barton (1998), Product sequences in azeotropic batch distillation, *AIChE J.*, **44**, 1051.

Anderson N. J. and M. F. Doherty (1984), An approximate model for binary azeotropic distillation design, *Chemical Engineering Science*, **39**, 11.

Barreto A., I. Rodriguez-Donis, V. Gerbaud, and X. Joulia (2011), Optimization of heterogeneous batch extractive distillation, to appear in *Ind. Eng. Chem. Res.*

Bernot C., M. F. Doherty, and M. F. Malone (1990), Patterns of composition changes in multicomponent batch distillation, *Chemical Engineering Science*, **45**, 1207.

Bernot C., M. F. Doherty, and M. F. Malone (1991), Feasibility and separation sequencing in multicomponent batch distillation, *Chemical Engineering Science*, **46**, 1311.

Bossen B. S., S. B. Jorgensen, and R. Gani (1993), Simulation, design, and analysis of azeotropic distillation, *Industrial and Engineering Chemistry Research*, **32**, 620.

Cheong W. and P. I. Barton (1999), Azeotropic distillation in a middle vessel batch column. 1. Model formulation and linear separation boundaries, *Ind. Eng. Chem. Res.*, **38**, 1504.

Cullie P. E. and G. V. Reklaitis (1986), Dynamic simulation of multicomponent batch rectification with chemical reactions, *Computers and Chemical Engineering*, **10**, 389.

Diwekar U. M. (1991), An efficient design method for binary azeotropic batch distillation, *AIChE Journal*, **14**, 946.

Doherty M. F. and M. F. Malone (2001), *Conceptual Design of Distillation Systems*, McGraw-Hill, New York.

Doherty M. F. and J. D. Perkins (1977), Properties of liquid-vapor composition surfaces at azeotropic points, *Chemical Engineering Science*, **32**, 112.

Doherty M. F. and J. D. Perkins (1978a), On the dynamics of distillation processes I. The simple distillation of multicomponent non-reacting, homogeneous liquid mixtures, *Chemical Engineering Science*, **33**, 281.

Doherty M. F. and J. D. Perkins (1978b), On the dynamics of distillation processes II. The simple distillation of model solutions, *Chemical Engineering Science*, **33**, 569.

Doherty M. F. and J. D. Perkins (1979), On the dynamics of distillation processes III. Topological classification of ternary residue curve maps, *Chemical Engineering Science*, **34**, 1401.

Ewell R. H. and L. M. Welch (1945), Rectification in ternary systems containing binary azeotropes, *Industrial Engineering Chemistry*, **37**, 1224.

Foucher, E. R., M. F. Doherty, and M. F. Malone (1991), Automatic screening of entrainers for homogeneous azeotropic distillation, *Industrial and Engineering Chemistry Research*, **29**, 760.

Gmehling J. and U. Onken (1982), *Vapor-Liquid Equilibrium Data Collection*, DECHEMA, Frankfurt.

Guo Z., M. Ghufran, and J. Lee (2003), Feasible products in batch reactive distillation, *AIChE Journal*, **49**, 3161.

Huerta-Garrido M., V. Rico-Ramirez, and S. Hernandez-Castro (2004), Simplified design of batch reactive distillation columns, *Ind. Eng. Chem. Res.*, **43**, 4000.

Kalagnanam J. R. (1991), Qualitative analysis of system behavior, PhD thesis, Department of Engineering & Public Policy, Carnegie Mellon University, Pittsburgh, PA.

Kalagnanam J. R. and U. M. Diwekar (1993), An application of qualitative analysis of ordinary differential equations to azeotropic batch distillation, *AI in Engineering*, **8**, 23.

Kim K.-J., U. M. Diwekar, and K. G. Tomazi (2004), Entrainer selection and solvent recycling in complex batch distillation, *Chemical Engineering Communications*,**191**, 1606.

Lelkes Z., P. Lang, B. Bendda and P. Moszkowicz (1998), Batch extractive distillation: The process and the operational policies, *AIChE J.*, **44**, 810.

Mujtaba I. (2004), *Batch Distillation Design and Operation*, Imperial College Press, London, United Kingdom.

Mujtaba I. and S. Macchietto (1997), Efficient optimization of batch distillation with chemical reaction using polynomial curve fitting techniques, *Ind. Eng. Chem. Res.*, **36**, 2287.

Pooling B., J. Prausnitz, and J. O'Connell (2001), *The Properties of Gases and Liquids*, fifth edition, McGraw-Hill, NY.

Rodriguez-Donis I., V. Gerbaud, and X. Joulia (2001a), Entrainer selection rules for the separation of azeotropic and close-boiling-temperature mixtures by homogeneous batch distillation process, *Ind. Eng. Chem. Res.*, **40**, 2729.

Rodriguez-Donis I., V. Gerbaud, and X. Joulia (2001b), Heterogeneous entrainer selection for the separation of azeotropic and close boiling temperature mixtures by heterogeneous batch distillation, *Ind. Eng. Chem. Res.*, **40**, 4935.

Rodriguez-Donis I., V. Gerbaud, and X. Joulia (2009a),Thermodynamic insights on the feasibility of homogeneous batch extractive distillation, 1. Azeotropic mixtures with a heavy entrainer, *Ind. Eng. Chem. Res.*, **48**, 3544.

Rodriguez-Donis I., V. Gerbaud, and X. Joulia (2009b), Thermodynamic insights on the feasibility of homogeneous batch extractive distillation, 2. Low-relative-volatility binary mixtures with a heavy entrainer, *Ind. Eng. Chem. Res.*, **48**, 3560.

Safrit B., A. W. Westerberg, U. M. Diwekar, and O. M. Wahnschafft (1995), Insights into batch extractive distillation using a middle vessel column, *Industrial and Engineering Chemistry Research*, **34**, 3257.

Safrit B. and Arthur W. Westerberg (1997), Improved operational policies for batch extractive distillation columns, *Ind. Eng. Chem. Res.*, **36**, 436.

Safrit, B. T. and A. W. Westerberg (1997a), Algorithm for generating the distillation regions for azeotropic multicomponent mixtures, *Ind. Chem. Eng. Res.*,**36**, 1827.

Safrit, B. T. and A. W. Westerberg (1997b), Synthesis of azeotropic batch distillation separation systems,*Ind. Chem. Eng. Res.*, **36**, 1841.

vanDongen D. B. and M. F. Doherty (1985a), Design and synthesis of homogeneous azeotropic distillations - I: Problem formulation for single column, *Industrial Engineering Chemistry Fundamentals*, **24**, 454.

vanDongen D. B. and M. F. Doherty (1985b), On the dynamics of distillation processes VI: Batch distillation, *Chemical Engineering Science*, **40**, 2087.

Venimadhavan G., G. Buzad, M. F. Doherty, and M. F. Malone (1994), Effect of kinetics on residue curve maps for reactive distillation, *AIChE J.*, **40**, 1814.

Exercises

7.1 Derive the residue curve map information for the acetone-benzene-chloroform azeotropic system using the thermodynamic data from Gmehling and Onken (1982).

7.2 Figure 7.45 shows the residue curve map for a hexane-methyl acetate-methanol system. Identify the still paths for batch distillation. Also describe how you would identify the batch distillation regions for this system.

7.3 The reaction of 2-toluene with no change in moles is given below. The data for the reaction and separation is given in Table 7.23. Find the profile of average distillate composition for all three components with time.

$$2 - toluene(C_7H_8) \Rightarrow benzene(C_6H_6) + o - xylene(C_8H_{10})$$

Vapor pressure is calculated in bar and temperature in degrees C using the following equation:

$$\log A - \frac{B}{T+C}$$

7.4 Present the degrees of freedom analysis for the McCabe–Thiele method for reactive batch distillation.

7.5 Table 7.24 (Mujtaba, 2004) presents the reactions and boiling points of reactants and products. Indicate which column you would use and why.

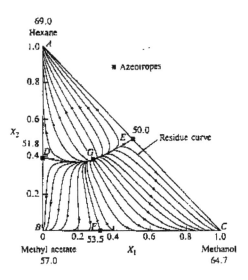

FIGURE 7.45
Residue Curve Map for the Hexane–Methyl Acetate–Methanol System (Reproduced from VanDongen and Doherty, 1985a)

TABLE 7.23

Data for the Multicomponent Reaction-Separation System

Parameter	Value
No. of Stages	2 +reboiler
Feed	100 $kmol$
Feed Comp. (Benzene)	0.006
Feed Comp. (O-Xylene)	0.485
Vapor Boilup Rate	140 $kmolh^{-1}$
Column Pressure	1 atm
Batch Time	1. h
Order of Reaction	1
Activation Energy	30,000 $kjkmol^{-1}$
Frequency Factor	8.0 E2 h^{-1}
Volumetric Flow Rate	2 m^3h^{-1}
Liquid Holdup	4 E-4 m^3
Antoine Constants:A,B,C (Pooling et al., 2001)	
benzene	3.98523,1184.240, 217.572
o-xylene	4.09789,1458.706, 212.041
2-toluene	4.05043, 1327.62, 217.625

TABLE 7.24

Find the Suitable Batch Distillation Column

$A + B \Rightarrow C$			
A (Boiling Point, K)	B (Boiling Point, K)	C (Boiling Point, K)	
Ethylene Oxide (283.5)	Water (373.2)	Ethylene Glycol (470.4)	
A (Boiling Point, K)	B (Boiling Point, K)	C (Boiling Point, K)	D (Boiling Point, K)
Acetic Acid (391.1)	Methanol (337.8)	Methyl Acetate (330.1)	Water (373.2)
Acetic Acid (391.1)	Ethanol (351.5)	Ethyl Acetate (350.3)	Water (373.2)
Acetic Acid (391.1)	Butanol (390.9)	Butyl Acetate (399.2)	Water (373.2)

8

BATCH DISTILLATION CONTROL

CONTENTS

Optimal control problems in batch distillation have received considerable attention in the literature. In general, control refers to a closed-loop system where the desired operating point is compared to an actual operating point and a knowledge of the difference (error) is fed back to the system to drive the actual operating point towards the desired one. The purpose of designing a closed-loop control system is to decrease the sensitivity of the plant to external disturbances. Conventional frequency domain techniques can be used to design a controller. However, optimal control problems do not fall under this category. These problems are defined in the time domain, and their solution requires establishing an index of performance for the system and designing the system so as to optimize the selected index. Use of the control function here provides an open-loop control. Since the decision variables that result in optimal performance are time-dependent, this type of control problem is referred to as an optimal control problem. The dynamic nature of these decision variables makes these problems much more difficult to solve compared to normal optimization where the decision variables are scalar.

The following section describes the different optimal control problems reported in the batch distillation literature and the optimal control techniques for solving these problems. This section is followed by a section on closed-loop control problems in batch distillation.

8.1   Optimal Control in Batch Distillation

As stated earlier, in the context of batch distillation, the optimization problems reported in the literature are mostly for the optimal operation of a column for a fixed design and are popularly referred to as optimal control problems. These problems can be classified as:

- Maximum Distillate Problem – where the amount of distillate of a specified concentration for a specified time is maximized (Converse and Gross, 1963; Keith and Brunet, 1971; Murty et al., 1980; Diwekar et al., 1987; Logsdon et al., 1990; Farhat et al., 1990; Diwekar, 1992; Logsdon and Biegler, 1993; Meski and Morari, 1995).

- Minimum Time Problem – where the batch time needed to produce a prescribed amount of distillate of a specified concentration is minimized (Coward, 1967a, b; Robinson, 1969, 1970; Mayur and Jackson, 1971; Egly et al., 1979; Hansen and Jorgensen, 1986; Christensen and Jorgensen, 1987; Mujtaba and Macchietto, 1988, 1992; Diwekar, 1992; Bonny et al., 1996; Mujtaba and Macchietto, 1996).

- Maximum Profit Problem – where a profit function for a specified concentration of distillate is maximized (Kerkhof and Vissers, 1978; Logsdon et al., 1990; Bonny et al., 1996; Li et al., 1997; Mujtaba and Macchietto, 1997; Wajge and Reklaitis, 1998; Hasebe et al., 1999).

The following paragraphs provide the mathematical representation of these problems. We begin with a brief review of the batch distillation model which will be used in the optimal control problems.

Batch Distillation Column Model

Most of the models used to solve optimal control problems in batch distillation neglect the holdup effect, thus justifying the use of quasi-steady state approximations to represent the column calculations. Coward (1967b), Robinson (1970), and Logsdon and Biegler (1993) have shown that the optimal control profiles obtained using negligible holdup effects differ from those using models with significant holdup. They observed that in some cases, the optimal reflux policy with the consideration of holdup has a period of zero reflux which is not observed in the optimal reflux policy obtained using the zero holdup models. However, for a binary system with a ten-plate column, Logsdon and Biegler (1993) had to solve a nonlinear programming problem (NLP) with 8000 decision variables, a very large NLP problem. Several factors must be considered before determining whether holdup effects alter the performance of a column, as discussed in Chapter 3. In this chapter, for the sake of simplicity, we restrict our discussions to negligible holdup models. However, the methods presented here could be easily extended to models with plate holdups also, provided an appropriate model is employed. For ease of understanding the basic concepts, the batch distillation model assuming quasi-steady state approximation for the column with the dynamics of the reboiler will be considered, which is also consistent with the many problems reported in the literature. Diwekar et al. (1987) suggested another simplification for multicomponent systems, wherein the quasi-steady state approximation extends to components in the reboiler other than the key component (in our discussions the key component is considered to be component 1). Figure 8.1 shows the schematic of the batch distillation model, for which, under the assumption of a constant boilup rate

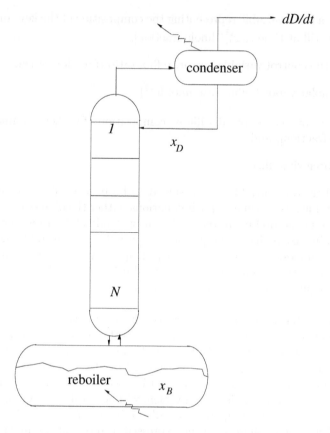

FIGURE 8.1
Schematic of a Batch Distillation Column

and no holdup conditions, an overall differential material balance equation over a time dt may be written as

$$\frac{dx_t^{(1)}}{dt} = \frac{dB_t}{dt} = \frac{-V}{R_t + 1}, \quad x_0^{(1)} = B_0 = F, \tag{8.1}$$

where F is the initial feed, at time $t = 0$, i.e., $x_t^{(1)}$. Similarly, a material balance for key component 1 over the differential time dt is

$$\frac{dx_t^{(2)}}{dt} = \frac{V}{R_t + 1}\frac{\left(x_B^{(1)} - x_D^{(1)}\right)}{B_t}, \quad x_0^{(2)} = x_F^{(1)} \tag{8.2}$$

In the above two equations

$x_t^{(1)}$ = a state variable at time t representing quantity of charge remaining in the still, B_t [mol],

$x_t^{(2)}$ = a state variable representing the composition of the key component in the still at time t, $x_B^{(1)}$ [mole fraction],

R_t = the control variable vector, reflux ratio (function of time),

V = molar vapor boilup rate [mol h^{-1}]

$x_D^{(1)}$ = overhead or distillate composition for key component 1 [mole fraction], and

t = batch time [hr].

The other state variables are assumed to be in a quasi-steady state and may be obtained using time-implicit equations (either the plate-to-plate model or the shortcut method equations). The quasi-steady state approximation reduces the dimensionality of the problem, especially if one is applying optimal control techniques, such as maximum principle, dynamic programming, or calculus of variations, since the number of additional variables and additional equations required for these techniques do not increase with the number of components.

As can be seen from the above discussion, batch distillation optimal control problems differ in the type of index of performance as well as solution techniques. The next section deals with the mathematical definition of the different optimal control problems categorized by the indices of performance. This is followed by a section on the solution techniques for optimal control problems. The maximum distillate problem will be considered for the purpose of illustration of the techniques, since all the indices for the optimal control of batch distillation columns can be expressed in terms of the maximum distillate problem (Diwekar, 1992). The feasibility considerations for the design variables have been discussed by Diwekar (1992) and are presented in Chapter 5. Please refer to Section 5.3 for details.

The Maximum Distillate Problem

Converse and Gross (1963) were the first to report the maximum distillate problem for binary batch distillation columns, which was solved using Pontryagin's maximum principle, the dynamic programming scheme, and the calculus of variations. Diwekar et al. (1987) extended the model for multicomponent systems and used the shortcut method along with the maximum principle for the calculation of optimal reflux policy. Logsdon et al. (1990) used finite element collocation and nonlinear programming optimization (NLP) techniques to solve the same problem using the shortcut model. The objective function used by Farhat et al. (1990) may be classified as a maximum distillate problem, although their aim was to maximize production of the specified cuts in multifraction operation. In their problem, the batch times for each cut were also considered as decision variables. They used linear and exponential approximations to the optimal profiles and applied NLP optimization techniques to obtain the solution. Recently, Meski and Morari (1995) presented the maximum distillate problem for the middle vessel column and proposed an optimal operating strategy.

The maximum distillate problem described in the literature as early as 1963 (Converse and Gross, 1963) can be represented as follows.

$$\text{Maximize} \quad J = \int_0^T \frac{dD}{dt}\, dt = \int_0^T \frac{V}{R_t + 1}\, dt, \qquad (8.3)$$
$$R_t$$

subject to the following purity constraint on the distillate

$$x_{Dave} = \frac{\int_0^T x_D^{(1)} \frac{V}{R_t + 1}\, dt}{\int_0^T \frac{V}{R_t + 1}\, dt} = x_D^* \qquad (8.4)$$

The problem can be rewritten as (from Equations 8.1 and 8.3)

$$\text{Maximize} \quad -x_T^{(1)} \qquad (8.5)$$
$$R_t$$

subject to Equations 8.1 and 8.2 and the time-implicit model of the column, which provides correlations between the model parameters and the state variables.

The Minimum Time Problem

Early approaches to the minimum time problem involved solution of two-point value problems (Coward, 1967a, b; Robinson, 1969; Mayur and Jackson, 1971; Egly et al., 1979; Hansen and Jorgensen, 1986). Mujtaba and Macchietto (1988) used piecewise constant reflux policy and the sequential quadratic programming (SQP) NLP optimizer to solve the problem. They extended their previous work to the optimal recycle policy for multicomponent systems, which were decomposed to a sequence of pseudo binary optimal control problems (Mujtaba and Macchietto, 1992).

There are several different formulations of the minimum time problem. Coward (1967a,b) converted the time-dependent equation, Equation 8.2, in terms of derivatives with respect to the amount remaining in the still B, and similarly changed the right hand side of Equation 8.1 to represent the derivative of time with respect to B. Since time is a dependent variable in this formulation, the minimum time problem could be easily stated. However, to establish the unified theory for all optimal control problems, Diwekar (1992) used the following formulation.

For the minimum time problem presented by Diwekar (1992), time-dependent Equations 8.1 and 8.2, and the time-implicit model equations remain the same. However, a new dummy variable t^* is introduced as the third state variable, and the relation of t^* with the actual batch time t is given by

$$\frac{dx_t^{(3)}}{dt} = \frac{dt^*}{dt} = 1, \quad t_0^* = 0 \tag{8.6}$$

The minimum time problem can then be written as:

$$\text{Minimize} \quad J = \int_0^T \frac{dt^*}{dt} \, dt, \tag{8.7}$$

$$R_t$$

subject to the purity constraint on the distillate:

$$x_{Dav} = \frac{\int_0^T x_D^{(1)} \frac{V}{R_t + 1} \, dt}{\int_0^T \frac{V}{R_t + 1} \, dt} = x_D^* \tag{8.8}$$

The Maximum Profit Problem

Very few researchers have solved the problem of optimizing a profit function for constant reflux, variable reflux, or optimal reflux operating conditions. Kerkhof and Vissers (1978), who were the first to use the profit function for maximization, solved the optimal control problem; however, their objective function did not include the effect of the number of plates and vapor boilup rate. Diwekar et al. (1987) used a different objective function to solve the profit maximization problem for constant and variable reflux conditions. This profit function is also presented in Chapter 5 in the context of design optimization. Logsdon et al. (1990) formulated a new profit function and solved the differential algebraic optimization problem (DAOP) for optimal design and operation. Bonny et al. (1996) investigated the optimal off-cut recycling and mixing strategies from the superstructure of batch distillation for maximum profit. Li et al. (1997) developed a detailed dynamic multifraction batch distillation model and discretized the model using the orthogonal collocation method on finite elements, and finally solved the maximum profit model using an NLP optimizer. Mujtaba and Macchietto (1997) considered a rigorous reactive distillation system for the maximum conversion problem, which can also be classified as the maximum profit problem. The detailed dynamic system is then reduced by using polynomial curve fitting techniques and solved by using an NLP optimizer. Wajge and Reklaitis (1998) also considered reactive batch distillation where the objective is to find an optimal campaign structure of the batch system with or without recycling.

A variant of this objective function is to minimize the mean rate of energy consumption when the market size for the product is fixed by the current demand. The objective function is given by Furlonge et al. (1999):

$$\text{Minimize} \quad J = \frac{\int_0^T Q_R(t)dt}{T + t_s},$$ (8.9)

$$\text{subject to} \quad x_{Davg} \geq x^*,$$
$$D \geq D^*$$

where Q_R is the reboiler heat duty. They used this objective function for optimal control of multivessel columns for the first time. Hasebe et al. (1999) also presented the optimal operation policy based on energy consumption for the multivessel column.

It has been shown that all the maximum profit problems can be written in terms of the maximum distillate problem (Diwekar, 1992), as explained below.

- Maximization of profit (Kerkhof and Vissers, 1978):

$$\text{Maximize} \quad J = \frac{DP_r - B_0C_0}{T + t_s}$$ (8.10)
$$R_t, T$$

subject to the purity constraint on the distillate:

$$x_{Dav} = \frac{\int_0^T x_D^{(1)} \frac{V}{R_t + 1} \, dt}{\int_0^T \frac{V}{R_t + 1} \, dt} = x_D^*$$ (8.11)

where

D = amount of product distilled [mol]

P_r = sales value of the product [\$/mole]

B_0 = amount of feed F [mol]

C_0 = cost of feed [\$/mole]

T = batch time [hr]

t_s = setup time for each batch [hr]

The problem may be converted as:

$$\text{Maximize} \quad \frac{(\overset{\text{Maximize}}{R_t} D) P_r - B_0C_0}{T + t_s}$$ (8.12)
$$T$$

- Maximization of profit (Diwekar et al., 1989):

$$\text{Maximize} \quad J = \frac{24(365)DP_r}{T + t_s} - \frac{c_1 VN}{G_a} - \frac{c_2 V}{G_b} - \frac{24(365)c_3 VT}{T + t_s}$$
$$R_t, N, T, V$$ (8.13)

where c_1, c_2, and c_3 are the cost coefficients, G_a is allowable vapor velocity, and G_b is the vapor handling capacity of the equipment. The objective function may be expressed as a maximum distillate problem, as shown below

$$\underset{N,T,V}{\text{Maximize}} \quad \frac{24(365)(\frac{\text{Maximize}}{R_t}D)P_r}{T+t_s} - \frac{c_1VN}{G_a} - \frac{c_2V}{G_b} - \frac{24(365)c_3VT}{T+t_s} \tag{8.14}$$

- Maximization of profit (Logsdon et al., 1990):

$$\underset{R_t,T,N}{\text{Maximize}} \quad J = \frac{DP_r - B_0C_0}{T+t_s} - \frac{K_1V^{0.5}N^{0.8} - K_2V^{0.65} - K_3V}{HRs} \tag{8.15}$$

where K_1, K_2, and K_3 represent the cost coefficients and HRs represents the hours per year. To convert the problem for application of the maximum distillate problem,

$$\underset{T,N}{\text{Maximize}} \quad (\frac{(\frac{\text{Maximize}}{R_t}D)P_r - B_0C_0}{T+t_s} - \frac{K_1V^{0.5}N^{0.8} - K_2V^{0.65} - K_3V}{HRs}) \tag{8.16}$$

Solution Techniques

Optimal control methods such as the calculus of variations was one of the first techniques in the field of optimization theory. Pontryagin's maximum principle and Bellman's dynamic programming principle emerged around 1956-1957. Optimal control problems in batch distillation were solved using these three techniques as early as 1963. Optimization involving both differential and algebraic equation models (DAOP) has been the focus of attention in recent literature. Most of the literature in this area concentrates on discretizing ODEs using the collocation approach (Cuthrell and Biegler, 1987; Logsdon et al., 1990) or approximating the control profile by polynomials (Akgiray and Heydeweiller, 1990; Farhat et al., 1990; Mujtaba and Macchietto, 1988) and applying NLP methods to solve the problem. For nonlinear models, the combination of collocation techniques and NLP optimization increases nonlinearities in the system, thereby increasing the possibility of multiple solutions and requiring good initial guesses. On the other hand, polynomial approximation methods depend on the crucial choice of the right type and order of polynomials for the control profile approximation. Diwekar (1992) proposed a combination of the maximum principle and an NLP optimization technique as an efficient alternative to solving optimal control problems in batch distillation.

A differential algebraic optimization problem (DAOP) in general can be stated as follows:

$$\text{Optimize} \quad J = j(\overline{x}_T) + \int_0^T k(\overline{x}_t, \theta_t, \overline{\mu})\, dt \qquad (8.17)$$
$$\theta_t, \overline{\mu}$$

subject to

$$\frac{d\overline{x}_t}{dt} = f(\overline{x}_t, \theta_t, \overline{\mu}) \qquad (8.18)$$

$$h(\overline{x}_t, \theta_t, \overline{\mu}) = 0 \qquad (8.19)$$

$$g(\overline{x}_t, \theta_t, \overline{\mu}) \leq 0 \qquad (8.20)$$

$$\overline{x}_0 = \overline{x}_{initial}$$

$$\theta(L) \leq \theta_t \leq \theta(U)$$

$$\overline{\mu}(L) \leq \overline{\mu} \leq \overline{\mu}(U)$$

where J is the objective function given by Equation 8.17, $\overline{x}_t$ is the state vector ($nx \times 1$ dimensional) at any time t, θ_t is the control vector, and $\overline{\mu}$ is the vector of the scalar variables. Equations 8.19 and 8.20 represent the equality (m_1 constraints) and inequality constraints (m_2 constraints, including bounds on the state variables), respectively (constituting a total of m constraints). $\theta(L)$ and $\overline{\mu}(L)$ represent the lower bounds on the set of control variables θ_t and the scalar variable μ, respectively, while $\theta(U)$, $\overline{\mu}(U)$ are the corresponding upper bounds.

In the absence of the scalar variable vector $\overline{\mu}$, a DAOP is equivalent to an optimal control problem. As stated earlier, there are different approaches to solving this problem, which are described briefly below.

Calculus of Variations

As seen in Chapter 5, the theory of optimization began with the calculus of variation, which is based on the vanishing of the first variation of a functional according to the theorem of minimum potential energy, leading to the Euler equations and natural boundary conditions. A functional is defined as a quantity or function that depends upon the entire course or path of one or more functions rather than on a number of scalar variables. Application of the minimum-energy principle involves the definition of stationary values for a function, or a set of functionals. In the above optimal control definition, the objective function J is a functional which depends upon the entire path from time equal to zero to time equal to T. Remember that we are neglecting the bounds $\theta(L)$ and $\theta(U)$ on the control variables and assuming that the scaler variables μ are fixed. Also, at the first part of the derivation the constraints are not included. To obtain the extremum value of J, the total differential of Equation 8.17 is equated to zero, as follows:

$$\int_0^T dJ = \int_0^T \left[\frac{\partial J}{\partial \theta} \delta\theta + \frac{\partial J}{\partial \theta'} \delta\theta' \right] dt = 0 \qquad (8.21)$$

The left hand side is called the first variation of the integral J. In order to eliminate the variations with respect to $\delta\theta'$, where $\theta' = d\theta/dt$, from the above equation, the second term is integrated by parts.

$$\int_0^T \frac{\partial J}{\partial \theta'} \delta\theta' dt = \left[\frac{\partial J}{\partial \theta'} \right]_0^T [d\theta]_0^T - \int_0^T \frac{d\left(\frac{\partial J}{\partial \theta'}\right)}{dt} \delta\theta dt \qquad (8.22)$$

By substituting Equation 8.22 in Equation 8.21 and imposing the boundary condition that $d\theta = 0$ at $t = 0$ and $t = T$, the following equation results:

$$J = \int_0^T \left[\frac{\partial J}{\partial \theta} - \frac{d\left(\frac{\partial J}{\partial \theta'}\right)}{dt} \right] \delta\theta dt \qquad (8.23)$$

The above integral must vanish for all admissible values of $\partial\theta$, which requires that the expression inside the brackets in Equation 8.23 be zero, i.e.,

$$\frac{\partial J}{\partial \theta} - \frac{d\left(\frac{\partial J}{\partial \theta'}\right)}{dt} = 0 \qquad (8.24)$$

The above differential equation is known as the Euler differential equation corresponding to the functional given in Equation 8.17. This, together with the boundary conditions, determines the function θ.

If the functional J is also constrained by equality constraints, then the application of the variational calculus leads to Euler–Lagrangian equations. In Euler–Lagrangian formulation, the objective function is augmented to include constraints, through the use of Lagrangian multipliers λ.

The following example demonstrates the application of variational calculus to the maximum distillate problem in batch distillation.

Example 8.1: The maximum distillate problem for separation of a binary mixture involves solution of the following problem:

$$\text{Maximize} \quad J = \int_0^T \frac{dD}{dt} dt = \int_0^T \frac{V}{R_t + 1} dt \qquad (8.25)$$

$$R_t$$

subject to the following purity constraint on the distillate

$$x_{Dav} = \frac{\int_0^T x_D^{(1)} \frac{V}{R_t + 1} dt}{\int_0^T \frac{V}{R_t + 1} dt} = x_D^* \qquad (8.26)$$

and the following differential equations governing the dynamics of the column. The rest of the column is assumed to be at quasi-steady state and to obey the plate-to-plate or the shortcut method calculation equations.

$$\frac{dx_t^{(1)}}{dt} = \frac{dB_t}{dt} = \frac{-V}{R_t + 1}, \quad x_0^{(1)} = B_0 = F \quad (8.27)$$

$$\frac{dx_t^{(2)}}{dt} = \frac{dx_B^{(1)}}{dt} = \frac{V}{R_t + 1} \frac{(x_t^{(2)} - x_D^{(1)})}{x_t^{(1)}}, \quad x_0^{(2)} = x_F^{(1)} \quad (8.28)$$

Formulate the maximum distillate problem using the calculus of variations.

Solution: Since this problem contains equality constraints, we need to use the Euler–Lagrangian formulation. First, all three equality constraints (Equations 8.26 to 8.28) are augmented to the objective function to form a new objective function L given by:

Maximize
$$\int_0^T L = \int_0^T \frac{V}{R_t + 1} \left[1 - \lambda(x_D^* - x_D^{(1)}) \right] -$$

$x^{(1)}$, x^0, R_t

$$\mu_1 \left[\frac{dx_t^{(1)}}{dt} - \frac{-V}{R_t + 1} \right] - \mu_2 \left[\frac{dx_t^{(2)}}{dt} - \frac{V}{R_t + 1} \frac{(x_B^{(1)} - x_D^{(1)})}{x_t^{(1)}} \right]$$

$$(8.29)$$

where λ is a scalar Lagrange multiplier and μ_i, $i = 1, 2$ are the Lagrangian multipliers as a function of time. Application of Euler differential equations leads to the following three Euler-Lagrange equations.

$$\frac{\partial L}{\partial x_t^{(1)}} - \frac{d(\frac{\partial L}{\partial x_t^{(1)'}})}{dt} = 0 \quad (\frac{\partial L}{\partial x_t^{(1)'}}) |_T = 0$$

$$\frac{d\mu_1}{dt} = \mu_2 \left[\frac{V}{R_t + 1} \frac{(x_t^{(2)} - x_D^{(1)})}{x_t^{(1)2}} \right] \quad \mu_1 |_T = 0 \quad (8.30)$$

$$\frac{\partial L}{\partial x_t^{(2)}} - \frac{d(\frac{\partial L}{\partial x_t^{(2)'}})}{dt} = 0 \quad (\frac{\partial L}{\partial x_t^{(2)'}}) |_T = 0$$

$$\frac{d\mu_2}{dt} = -\frac{V}{R_t + 1} \lambda \left(\frac{\partial x_D^{(1)}}{\partial x_t^{(2)}} \right)_{R_t} - \mu_2 \frac{V}{x_t^{(1)}(R_t + 1)} \quad (8.31)$$

$$\left[1 - \left(\frac{\partial x_D^{(1)}}{\partial x_t^{(2)}} \right)_{R_t} \right] \quad \mu_2 |_T = 0$$

$$(8.32)$$

$$\frac{\partial L}{\partial R_t} - \frac{d(\frac{\partial L}{\partial R_t'})}{dt} = 0$$

$$R_t = \frac{\left[\frac{\mu_2}{x_t^{(1)}}(x_t^{(2)} - x_D^{(1)}) - \mu_1 - \lambda(x_D^* - x_D^{(1)}) + 1 \right]}{\frac{\partial x_D^{(1)}}{\partial R_t} \left(\lambda - \frac{\mu_2}{x_t^{(1)}} \right) - 1} \quad (8.33)$$

The Maximum Principle

The maximum principle was first proposed in 1956 by Pontryagin and co-workers (Boltyanskii et al., 1956; Pontryagin, 1956, 1957). Since then it has been widely used to solve a variety of optimal control problems. Like variational calculus, the maximum principle can be used to solve an optimal control problem for a fixed scalar variable vector $(\overline{\mu})$ only, and not the complete DAOP described in the previous section.

In the maximum principle formulation (the right hand side of Equation 8.34) the objective function is represented as a linear function in terms of the final values of $\overline{x}$ and the values of $\overline{c}$, where $\overline{c}$ represents the vector of constants. The maximum principle formulation for the above mentioned DAOP is given below.

$$\text{Maximize} \quad J = j(\overline{x}_T) + \int_0^T k(\overline{x}_t, \theta_t, \overline{\mu} \; dt) = \overline{c}^T \overline{x}_T = \sum_{i=1}^{nx} c^{(i)} x_T^{(i)}$$
$$\theta_t \tag{8.34}$$

subject to

$$\frac{d\overline{x}_t}{dt} = f(\overline{x}_t, \theta_t, \overline{\mu}) \tag{8.35}$$

$$h(\overline{x}_t, \theta_t, \overline{\mu}) = 0 \tag{8.36}$$

$$g(\overline{x}_t, \theta_t, \overline{\mu}) \leq 0 \tag{8.37}$$

$$\overline{x}_0 = \overline{x}_{initial}$$

$$\theta(L) \leq \theta_t \leq \theta(U)$$

By using the Lagrangian formulation for the above problem and by removing the bounds $\theta(L)$ and $\theta(U)$ on the control variable vector θ_t, since the maximum principle cannot easily handle the bounds on the control variable (Cuthrell and Biegler, 1987; Akgiray and Heydeweiller, 1990), one obtains:

$$\text{Maximize} \quad J^* = \overline{c}^T \overline{x}_T + \overline{\lambda_1}^T (h(\overline{x}_t, \theta_t, \overline{\mu})) + \overline{\lambda_2}^T (g(\overline{x}_t, \theta_t, \overline{\mu}))$$
$$\theta_t \tag{8.38}$$

subject to

$$\frac{d\overline{x}_t}{dt} = f(\overline{x}_t, \theta_t, \overline{\mu}) \tag{8.39}$$

$$\overline{x}_0 = \overline{x}_{initial}$$

where

$$\overline{\lambda}^T = [\overline{\lambda_1}^T, \overline{\lambda_2}^T]$$

Application of the maximum principle to the above problem involves the addition of nx adjoint variables z_t (one adjoint variable per state variable), nx adjoint equations, and a Hamiltonian, which satisfies the following relations:

$$H(\overline{z}_t, \overline{x}_t, \theta_t) \;=\; \overline{z}_t^T \, f(\overline{x}_t, \theta_t) \;=\; \sum_{i=1}^{nx} z_t^{(i)} f^{(i)}(\overline{x}_t, \theta_t) \tag{8.40}$$

$$\frac{dz_t^{(i)}}{dt} \;=\; -\sum_{j=1}^{n} z_t^{(j)} \frac{\partial f^{(j)}}{\partial x_t^i} \tag{8.41}$$

$$\overline{z}_T \;=\; \overline{c}$$

The optimal decision vector θ_t can be obtained by extremizing the Hamiltonian given by Equation 8.40. θ_t can then be expressed as

$$\theta_t \;=\; H^*(\overline{x}_t, \overline{z}_t, \overline{\lambda}) \tag{8.42}$$

It is possible to derive the necessary condition for optimality in the calculus of variations from the maximum principle when the decision vector is not constrained. Conversely, by using the technique of calculus of variation the weakened form of the maximum principle can be derived. These derivations are presented in Fan (1966) and the interested reader is referred to this book on the maximum principle for further details.

Example 8.2: Formulate the maximum distillate problem presented in Example 8.1 using the maximum principle and compare the formulation with that obtained using variational calculus (solution for Example 8.1).

Solution: The maximum distillate problem can be written as

$$\text{Maximize} \quad J = \int_0^T \frac{dD}{dt}\, dt = \int_0^T \frac{V}{R_t + 1}\, dt, \tag{8.43}$$
$$R_t$$

subject to the following purity constraint on the distillate

$$x_{Dav} = \frac{\int_0^T x_D^{(1)} \frac{V}{R_t + 1}\, dt}{\int_0^T \frac{V}{R_t + 1}\, dt} = x_D^* \tag{8.44}$$

The constraint on the purity is removed by employing the method of Lagrange multipliers. By combining Equations 8.43 and 8.44:

$$\text{Maximize} \quad L = \int_0^T \frac{V}{R_t + 1}\left[1 - \lambda(x_D^* - x_D^{(1)})\right] dt \tag{8.45}$$
$$R_t$$

where λ is a Lagrange multiplier. Now the objective function is to maximize L instead of J. To solve this problem, an additional state variable x_3 is introduced, which is given by

$$x_t^{(3)} = \int_0^t \frac{V}{R_t + 1}\left[1 - \lambda(x_D^* - x_D^{(1)})\right] dt \qquad (8.46)$$

The problem can then be rewritten as

$$\text{Maximize} \quad x_T^{(3)} \qquad (8.47)$$
$$R_t$$

subject to the following differential equations for the three state variables and the time-implicit model for the rest of the column,

$$\frac{dx_t^{(1)}}{dt} = \frac{-V}{R_t + 1}, \quad x_0^{(1)} = B_0 = F \qquad (8.48)$$

$$\frac{(x_t^{(2)} - x_D^{(1)})}{x_t^{(1)}}, \quad x_0^{(2)} = x_F^{(1)} \qquad (8.49)$$

$$\frac{dx_t^{(3)}}{dt} = \frac{V}{R_t + 1}\left[1 - \lambda(x_D^* - x_D^{(1)})\right] \qquad (8.50)$$

The Hamiltonian function, which should be maximized, is:

$$H_t = -z_t^{(1)}\frac{V}{R_t + 1} + z_t^{(2)}\frac{V(x_t^{(2)} - x_D^{(1)})}{(R_t + 1)x_t^{(1)}} + z_t^{(3)}\frac{V}{R_t + 1}\left[1 - \lambda(x_D^* - x_D^{(1)})\right]$$
$$(8.51)$$

and the adjoint equations are:

$$\frac{dz_t^{(1)}}{dt} = z_t^{(2)}\frac{V(x_t^{(2)} - x_D^{(1)})}{(R_t + 1)(x_t^{(1)})^2}, \quad z_T^{(1)} = 0 \qquad (8.52)$$

$$\frac{dz_t^{(2)}}{dt} = -z_t^{(2)}\frac{V(1 - \frac{\partial x_D^{(1)}}{\partial x_t^{(2)}})}{(R_t + 1)x_t^{(1)}} - z_t^{(3)}\lambda\frac{V}{(R_t + 1)}\left(\frac{\partial x_D^{(1)}}{\partial x_t^{(2)}}\right) \quad z_T^{(2)} = 0 \qquad (8.53)$$

and

$$\frac{dz_t^{(3)}}{dt} = 0, \quad z_T^{(3)} = 1 \qquad (8.54)$$

Since the above equation for z_t^3 gives

$$z_t^{(3)} = 1 \qquad (8.55)$$

the Hamiltonian function in Equation 8.51 can be written as

$$H_t = -z_t^{(1)}\frac{V}{R_t + 1} + z_t^{(2)}\frac{V(x_t^{(2)} - x_D^{(1)})}{(R_t + 1)x_t^{(1)}} + \frac{V}{R_t + 1}\left[1 - \lambda(x_D^* - x_D^{(1)})\right]$$
$$(8.56)$$

and

$$\frac{dz_t^{(2)}}{dt} = -z_t^{(2)}\frac{V\left(1 - \frac{\partial x_D^{(1)}}{\partial x_t^{(2)}}\right)}{(R_t + 1)x_t^{(1)}} - \lambda\frac{V}{(R_t + 1)}\left(\frac{\partial x_D^{(1)}}{\partial x_t^{(2)}}\right), \quad z_T^{(2)} = 0 \qquad (8.57)$$

From the optimality condition $\partial H/\partial R_t = 0$, it follows that

$$R_t = \frac{\left[\frac{z_t^{(2)}}{x_t^1}(x_t^{(2)} - x_D^{(1)}) - z_t^{(1)} - \lambda(x_D^* - x_D^{(1)}) + 1 \right]}{\frac{\partial x_D^{(1)}}{\partial R_t}\left(\lambda - \frac{z_t^{(2)}}{x_t^{(1)}} \right)} - 1 \qquad (8.58)$$

It can be easily seen from Equation 8.33 in Example 8.1 and from Equation 8.58 that the two formulations lead to the same results where in the case of variational calculus, the time-dependent Lagrange multipliers μ_i are equivalent to the adjoint variables $z^{(i)}$ in the maximum principle formulation.

Dynamic Programming

The method of dynamic programming is based on the principle of optimality, as stated by Bellman (1957).

An optimal policy has the property that whatever the initial state and initial decision are the remaining decisions must constitute an optimal policy with regard to the state resulting from the first decision.

In short, the principle of optimality states that the minimum value of a function is a function of the initial state and the initial time. Dynamic programming is best suited for multistage processes. However, the application of the dynamic programming technique to a continuously operating system leads to nonlinear partial differential equations—the Hamilton–Jacobi–Bellman (H-J-B) equation. A brief derivation of this equation is given below. For details, please refer to Bellman (1957), Aris (1961), and Kirk (1970).

The optimal control problem involves the process described by the state equations:

$$\frac{d\overline{x}_t}{dt} = f(\overline{x}_t, \theta_t, \overline{\mu}) \qquad (8.59)$$

to be controlled so as to minimize the performance measure given by J:

$$\text{Optimize} \quad J = j(\overline{x}_T) + \int_0^T k(\overline{x}_t, \theta_t)\, dt \qquad (8.60)$$
$$\theta_t$$

Introducing a dummy variable of integration τ, where $t \leq \tau \leq T$, the performance measure in the interval $[t, T]$ is:

$$\text{Optimize} \quad j(\overline{x}_T) + \int_t^T k(\overline{x}_\tau, \theta_\tau)\, d\tau \qquad (8.61)$$
$$\theta_\tau$$

By subdividing the interval, we obtain

$$\text{Optimize} \quad j(\overline{x}_T) \; + \; \int_{t}^{t+\triangle t} k(\overline{x}_\tau, \theta_\tau) \, d\tau \; + \; \int_{t+\triangle t}^{T} k(\overline{x}_\tau, \theta_\tau) \, d\tau$$

$$\theta_\tau \tag{8.62}$$

The principle of optimality requires that $J_{opt}(\overline{x}(t), \; t)$ is equal to

$$\text{Optimize} \quad J_{opt}(\overline{x}(t+\triangle t), \; t+\triangle t) \; + \; \int_{t}^{t+\triangle t} k(\overline{x}_\tau, \theta_\tau) \, d\tau$$

$$\theta_\tau \tag{8.63}$$

where $J_{opt}(\overline{x}(t+\triangle t), \; t+\triangle t)$ is the optimum cost of the process for the time interval $t + \triangle t \leq \tau \leq T$ with the initial state $\overline{x}(t+\triangle t)$.

Assuming that the second partial derivative of the function J exists and is bounded, we can expand $J_{opt}(\overline{x}(t+\triangle t), \; t+\triangle t)$ as a Taylor series (neglecting the higher derivatives) about the point $(\overline{x}(t), t)$ to obtain:

$$J_{opt}(\overline{x}(t), \; t) \;=\; \text{Optimize} \int_{t}^{t+\triangle t} k(\overline{x}_\tau, \theta_\tau) \, d\tau \; + \; J(\overline{x}(t), \; t)$$

$$\theta_\tau$$

$$+ \quad \left[\frac{\partial J}{\partial t}\right] \triangle t \; + \; \left[\frac{\partial J}{\partial \overline{x}}\right]^\tau [x(t+\triangle t) \; - \; x(t)]$$

$$\tag{8.64}$$

For small $\triangle t$ the above equation reduces to:

$$J_{opt}(\overline{x}(t), \; t) \;=\; \text{Optimize} \quad k(\overline{x}_t, \theta_t) \triangle t \; + \; J(\overline{x}(t), \; t))$$

$$\theta_t$$

$$+ \quad \left[\frac{\partial J}{\partial t}\right] \triangle t \; + \; \left[\frac{\partial J}{\partial \overline{x}}\right]^\tau [dx(t)] \tag{8.65}$$

Dividing Equation 8.65 by $\triangle t$ and substituting the value of dx/dt from the differential equation 8.59 and further by virtue of the fact that the left hand side is not a function of θ_t, the following equation results

$$0 \;=\; \text{Optimize} \quad k(\overline{x}_t, \theta_t) \; + \; \left[\frac{\partial J}{\partial t}\right] \; + \; \left[\frac{\partial J}{\partial \overline{x}}\right]^\tau [f(\overline{x}_t, \theta_t, \overline{\mu}]$$

$$\theta_t \tag{8.66}$$

Defining the Hamiltonian as a function of $\overline{x}(t)$, $\partial J/\partial \overline{x}$, t, the above equation results in what is referred to as the Hamilton–Jacobi–Bellman equation,

$$0 \;=\; \left[\frac{\partial J}{\partial t}\right] \; + \; H\left(\overline{x}(t), \frac{\partial J}{\partial \overline{x}}, t\right) \tag{8.67}$$

where

$$H = \begin{array}{c} \text{Optimize} \\ \theta_T \end{array} k(\overline{x}_T, \theta_T) + \left[\frac{\partial J}{\partial \overline{x}}\right]^T [f(\overline{x}_T, \theta_T, \overline{\mu}]$$

(8.68)

With the boundary conditions:

$$\frac{\partial J}{\partial x(0)} = 0$$

(8.69)

$$\left[\frac{\partial J}{\partial t}\right]^T = H(\overline{x}(T), \frac{\partial J}{\partial \overline{x}}, T)$$

(8.70)

As can be seen, the H-J-B equation, a partial differential equation, is tedious to solve. Aris (1961) used the method of characteristics, described below for the above equations. In the method of characteristics a dummy variable s is introduced which converts the above equations into the series of equations given below.

$$\frac{dT}{ds} = 1, \quad T_{initial} = 0$$

(8.71)

$$\frac{d\overline{x}(T)}{ds} = f(\overline{x}_T, \theta_T, \overline{\mu})$$

(8.72)

$$\frac{dJ}{ds} = k(\overline{x}_T, \theta_T)$$

(8.73)

$$\frac{d\left(\frac{\partial J}{\partial x_T^{(i)}}\right)}{ds} = J(x^{(i)}(T), \ T) - \sum \left(\frac{\partial J}{\partial x_T^{(i)}}\right) f(x_T^{(i)}, \theta_T, \overline{\mu})$$

$$\left(\frac{\partial J}{\partial x_T^{(i)}}\right)_0 = 0$$

(8.74)

$$\frac{d\left(\frac{\partial J}{\partial T}\right)}{ds} = 0$$

$$\left(\frac{\partial J}{\partial T}\right)_0 = 0$$

(8.75)

These equations along with the boundary conditions need to be solved to get the optimal control policy.

Example 8.3: Solve the maximum distillate problem presented in Example 8.1 using the dynamic programming technique and compare the formulation with that obtained using variational calculus (solution for Example 8.1) and the maximum principle (solution for Example 8.2).

Solution: Once again the Lagrange multiplier formulation of the maximum distillate problem can be written as

$$\text{Maximize} \quad \int_0^T L = \int_0^T \frac{V}{R_t + 1}\left[1 - \lambda(x_D^* - x_D^{(1)})\right] dt \quad (8.76)$$
$$R_t$$

subject to:

$$\frac{dx_t^{(1)}}{dt} = \frac{-V}{R_t + 1}, \quad x_0^{(1)} = B_0 = F \quad (8.77)$$

$$\frac{dx_t^{(2)}}{dt} = \frac{V}{R_t + 1}\frac{(x_t^{(2)} - x_D^{(1)})}{x_t^{(1)}}, \quad x_0^{(2)} = x_F^{(1)} \quad (8.78)$$

The Hamilton–Jacobi–Bellman equation for this problem can be written as:

$$\frac{\partial L}{\partial t} + \begin{array}{c}\text{Optimize}\\ R_T\end{array} \frac{V}{R_T + 1}\frac{(x_T^{(2)} - x_D^{(1)})}{x_T^{(1)}}\frac{\partial L}{\partial x_T^{(2)}} + \frac{-V}{R_t + 1}\frac{\partial L}{\partial x_T^{(1)}} = 0$$
$$(8.79)$$

The above quasi-linear partial differential equation can be solved by using equations based on the method of characteristics (Equations 8.71-8.75).

$$\frac{dT}{ds} = 1, T_{initial} = 0 \quad (8.80)$$

$$\frac{dx_T^{(1)}}{ds} = \frac{-V}{R_T + 1} \quad (8.81)$$

$$\frac{dx_T^{(2)}}{ds} = \frac{V}{R_T + 1}\frac{(x_T^{(2)} - x_D^{(1)})}{x_T^{(1)}} \quad (8.82)$$

$$\frac{dL}{ds} = \frac{V}{R_T + 1}\left[1 + \lambda(x_D^{(1)} - x_D^*)\right] \quad (8.83)$$

$$\frac{d\left(\frac{\partial L}{\partial x_T^{(1)}}\right)}{ds} = \frac{-V}{R_T + 1}\frac{(x_T^{(2)} - x_D^{(1)})}{(x_T^{(1)})^2}\left(\frac{\partial L}{\partial x_T^{(2)}}\right)$$

$$\left(\frac{\partial L}{\partial x_T^{(1)}}\right)_T = 0 \quad (8.84)$$

$$\frac{d\left(\frac{\partial L}{\partial x_T^{(2)}}\right)}{ds} = \frac{V}{R_T + 1}\frac{\left(1 - \frac{\partial x_D^{(1)}}{\partial x_T^{(2)}}\right)}{(x_T^{(1)})}\left(\frac{\partial L}{\partial x_T^{(2)}}\right) + \frac{\lambda V}{R_T + 1}\left(\frac{\partial x_D^{(1)}}{\partial x_T^{(2)}}\right)_{R_T}$$

$$\left(\frac{\partial L}{\partial x_T^{(2)}}\right)_T = 0 \quad (8.85)$$

$$\frac{d\left(\frac{\partial L}{\partial T}\right)}{ds} = 0$$

$$\left(\frac{\partial L}{\partial T}\right)_T = H \quad (8.86)$$

$$H = \begin{array}{c}\text{Optimize}\\ R_T\end{array} \frac{V}{R_T + 1}\frac{(x_T^{(2)} - x_D^{(1)})}{x_T^{(1)}}\frac{\partial L}{\partial x_T^{(2)}} + \frac{-V}{R_T + 1}\frac{\partial L}{\partial x_T^{(1)}}$$

A differential equation for dR_T/ds could have been used along with the differential Equations 8.80 to 8.85 to obtain the optimal policy. However, an explicit equation can be obtained by differentiating the right hand side of Equation 8.86 with respect to R_T and setting it to zero, as given below.

$$R_T = \frac{\left[\frac{\partial L/\partial x_T^{(2)}}{x_T^{(1)}} (x_T^{(2)} - x_D^{(1)}) - \partial L/\partial x_T^{(1)} - \lambda(x_D^* - x_D^{(1)}) + 1 \right]}{\frac{\partial x_D^{(1)}}{\partial R_T} \left(\lambda - \frac{\partial L/\partial x_T^{(2)}}{x_T^{(1)}} \right)} - 1 \quad (8.87)$$

It can be easily seen from Equation 8.33 in Example 8.1 and Equation 8.58 in Example 8.2 as well as in the above equation that the formulations using dynamic programming, the maximum principle, and the calculus of variations lead to the same results, where, in the case of variational calculus, the time-dependent Lagrange multipliers μ_i are equivalent to the adjoint variables $z^{(i)}$ in the maximum principle formulation, which are equivalent to the partial derivatives of the function L with respect to the state variables in the dynamic programming formulation.

We have examined three optimal control methods to solve the maximum distillate problem in batch distillation. The three different methods gave the same results. However, to obtain the exact solution one has to solve this problem iteratively, which can be accomplished using several different methods, including the shooting method and the method of steepest ascent of the Hamiltonian. Although the details of the methods are beyond the scope of this book, the typical computational intensity involved in solving these problems is illustrated using the method of steepest ascent of the Hamiltonian (Diwekar et al., 1987) in the following example.

Example 8.4: Describe the solution procedure to solve the maximum distillate problem presented in Example 8.2 using the maximum principle formulation. Use the quasi-steady state shortcut method to model the batch distillation column.
Solution: The maximum principle formulation of the maximum distillate problem presented in Example 8.2 involves solution of the following equations:
The state variable differential equations.

$$\frac{dx_t^{(1)}}{dt} = \frac{-V}{R_t + 1}, \quad x_0^{(1)} = B_0 = F \quad (8.88)$$

$$\frac{dx_t^{(2)}}{dt} = \frac{V}{R_t + 1} \frac{(x_t^{(2)} - x_D^{(1)})}{x_t^{(1)}}, \quad x_0^{(2)} = x_F^{(1)} \quad (8.89)$$

The adjoint equations:

$$\frac{dz_t^{(1)}}{dt} = z_t^{(2)} \frac{V(x_t^{(2)} - x_D^{(1)})}{(R_t + 1)(x_t^{(1)})^2}, \quad z_T^{(1)} = 0 \quad (8.90)$$

$$\frac{dz_t^{(2)}}{dt} = -z_t^{(2)} \frac{V\left(1 - \frac{\partial x_D^{(1)}}{\partial x_t^{(2)}}\right)}{(R_t + 1)x_t^{(1)}} - \lambda \frac{V}{(R_t + 1)} \left(\frac{\partial x_D^{(1)}}{\partial x_t^{(2)}}\right), \quad z_T^{(2)} = 0 \quad (8.91)$$

The optimality conditions resulting in the following reflux ratio profile:

$$R_t = \frac{\left[\frac{z_t^{(2)}}{x_t^{(1)}} (x_t^{(2)} - x_D^{(1)}) - z_t^{(1)} - \lambda(x_D^* - x_D^{(1)}) + 1 \right]}{\frac{\partial x_D^{(1)}}{\partial R_t} \left(\lambda - \frac{z_t^{(2)}}{x_t^{(1)}} \right)} - 1 \tag{8.92}$$

where the relation between the state variables and the distillate composition, and the values of the partial derivatives $(\partial x_D^{(1)}/\partial R_t$ and $\partial x_D^{(1)}/\partial x_t^{(2)})$ can be obtained by the time-implicit shortcut simulation model presented below.

Shortcut simulation model

The equations used in the shortcut model presented in Chapter 4 are:
Hengestebeck–Geddes equation:

$$x_D^{(i)} = \left(\frac{\alpha_i}{\alpha_1} \right)^{C_1} \frac{x_D^{(1)}}{x_B^{(1)}} x_B^{(i)}, i = 2, 3, \ldots, n \tag{8.93}$$

where C_1 is equivalent to Fenske's N_{min}.
Underwood equations:

$$\sum_{i=1}^{n} \frac{\alpha_i x_B^{(i)}}{\alpha_i - \phi} = 0 \; ; \; R_{min\ u} + 1 = \sum_{i=1}^{n} \frac{\alpha_i x_D^{(i)}}{\alpha_i - \phi} \tag{8.94}$$

Gilliland's correlation:

$$Y = 1 - \exp[\frac{(1 + 54.4X)(X - 1)}{(11 + 117.2X)\sqrt{X}}] \tag{8.95}$$

in which

$$X = \frac{R - R_{min}}{R + 1}; \quad Y = \frac{N - N_{min}}{N + 1}$$

and

$$\sum_{i}^{n} x_D^{(i)} = 1 \tag{8.96}$$

In order to match the R_{min} predicted by the Underwood equations (where the composition $x_D^{(i)}$'s are used to calculate this R_{min}) with the R_{min} obtained by Gilliland's correlation (which uses N, R, and N_{min} or equivalently C_1), the following additional condition is provided:

$$G_c = \frac{R_{min\ g}}{R} - \frac{R_{min\ u}}{R} = 0 \tag{8.97}$$

For a specific value of the state variables at each time step, it is necessary to estimate the values of $x_D^{(i)}$, $i = 1, 2, \ldots, n$, and $x_B^{(i)}$, $i = 2, 3, \ldots, n$. Using the quasi-steady state approximation, the following differential equation can be used to obtain the $x_B^{(i)}$, $i = 2, 3, \ldots, n$ using the previous time step values.

$$x_{B_{new}}^{(i)} = x_{B_{old}}^{(i)} + \frac{\triangle x_t^{(2)} \left(x_D^{(i)} - x_B^{(i)} \right)}{\left(x_D^{(1)} - x_t^{(2)} \right)_{old}} \bigg|_{old}, i = 2, 3, \ldots, n \tag{8.98}$$

Assuming C_1 and substituting the Hengestebeck–Geddes equation in Equation 8.96 results in

$$x_D^{(1)} = \frac{1}{\sum_{i=1}^n \left(\frac{\alpha_i}{\alpha_1}\right)^{C_1} \frac{x_B^{(i)}}{x_t^{(2)}}} \tag{8.99}$$

Substituting Equation 8.99 in Equation 8.93 and then using Equation 8.97, one can obtain the value of C_1 and hence $x_D^{(i)}$, $i = 1, 2, \ldots, n$. The Newton-Raphson iterative procedure is used for the solution, the derivative required for which is given by:

$$
\begin{aligned}
R\frac{\partial G_c}{\partial C_1} &= -\sum_{i=1}^n \alpha_i (\alpha_i/\alpha_1)^{C_1} \frac{x_B^{(i)}}{x_t^{(2)}} \frac{\sum_{j=1}^n \left(\frac{\alpha_j}{\alpha_1}\right)^{C_1} \frac{x_B^{(i)}}{x_t^{(2)}} \ln\left(\frac{\alpha_j}{\alpha_1}\right)}{\sum_{j=1}^n \left[\left(\frac{\alpha_j}{\alpha_1}\right)^{C_1} \frac{x_B^{(i)}}{x_t^{(2)}}\right]^2} (\alpha_i - \phi) \\
&+ \sum_{i=1}^n \alpha_i \frac{x_D^{(i)} \ln(\alpha_i/\alpha_1)}{\alpha_i - \phi} - \frac{R+1}{(N+1)dY/dX}
\end{aligned} \tag{8.100}
$$

Similarly the values of $dx_D(1)/dR_t$ required for the solution can be obtained as follows:

$$
\begin{aligned}
\frac{dx_D(1)}{dR_t} &= \frac{dx_D(1)}{dC_1}\frac{dC_1}{dR} \tag{8.101} \\
&= \frac{\sum_{j=1}^n \left(\frac{\alpha_j}{\alpha_1}\right)^{C_1} \frac{x_B^{(i)}}{x_t^{(2)}} \ln\left(\frac{\alpha_j}{\alpha_1}\right)}{\sum_{j=1}^n \left[\left(\frac{\alpha_j}{\alpha_1}\right)^{C_1} \frac{x_B^{(i)}}{x_t^{(2)}}\right]^2} \\
&\quad \left[\frac{(R_{min\ g} - R_{min\ u})}{R^2} - \frac{(R_{min\ g} + 1)}{(R+1)R}\right] / \frac{dG_c}{dC_1} \tag{8.102}
\end{aligned}
$$

Basically now the problem is reduced to finding the solution of Equation 8.92 using Equations 8.88 to 8.91 along with the shortcut method equations. The equations also involve the Lagrange multiplier λ, which is a constant for a specific value of final time T. So, the above equations must be solved for different values of λ, until the purity constraint given below is satisfied.

$$x_{Dav} = \frac{\int_0^T x_D^{(1)} \frac{V}{R_t + 1}\, dt}{\int_0^T \frac{V}{R_t + 1}\, dt} = x_D^* \tag{8.103}$$

It can be seen that the solution of these equations also involves a two-point boundary value problem, where the initial values of the state variables $x_0^{(1)}$ and $x_0^{()}$ and the final values of the adjoint variables $z_T^{(1)}$ and $z_T^{(2)}$ are known. We seek the maximum value of H by choosing the decision vector R_t. The method of steepest ascent of the Hamiltonian accomplishes this by using an iterative procedure to find R_t, the optimal decision vector. An initial estimate of R_t is obtained, which is updated during each iteration. If the decision vector R_t is divided into r time intervals, then for the i-th component of the decision vector, the following rule is used for proceeding from the j-th to $j+1$-th approximation:

$$R_i(j+1) = R_i(j) + k\frac{\partial H}{\partial R_i}, i = 1, 2, \dots, r \qquad (8.104)$$

where k is a suitable constant. The iterative method is used until there is no further change in R_t. The value of k should be small enough so that no instability will result, yet large enough for rapid convergence. It should be noted that the sign of k is important, because $\partial H/\partial R_t \to 0$ at and near the total reflux condition, which gives the minimum value of H (i.e., minimum distillate).

Also, one has to iterate on the Lagrange multiplier in the outer loop so that the purity constraint is satisfied. Figure 8.2 describes this iterative procedure.

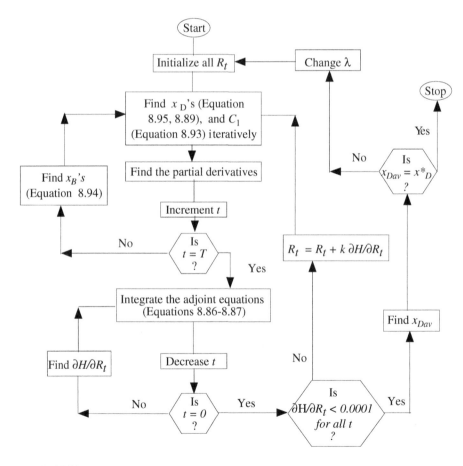

FIGURE 8.2
Flowchart for Example 8.4

NLP Optimization with Orthogonal Collocation

Chapter 4 introduced the collocation approach for solving differential equations, which involves discretization of the state variables in terms of polynomial approximation using the orthogonal collocation technique. The discretization converts the set of differential equations to a set of nonlinear algebraic equations, which can then be solved using NLP optimization techniques.

The approach is presented briefly for the general DAOP mentioned earlier. Discretization of the state and control variables in terms of Lagrange polynomials ($l_i(t)$) of *ncol* order (*ncol* is the number of collocation points) results in the following

$$\overline{x}_t = \sum_{i=1}^{ncol+2} l_i(t)\overline{x}_i; \; \theta_t = \sum_{i=1}^{ncol+2} l_i(t)\theta_i \qquad (8.105)$$

$$\frac{d\overline{x}_t}{dt} = \sum_{i=1}^{ncol+2} A_{ki}\overline{x}_i \qquad (8.106)$$

where A_{ki} are the coefficients arising from differentiating the Lagrange polynomials.

Substituting the values of $\overline{x}_i$, θ_i, and $d\overline{x}_i/dt$ from the above equations, the DAOP reduces to

$$\text{Optimize} \qquad j\left(\sum_{i=1}^{ncol+2} l_i(T)x_i + \sum_{i=1}^{ncol+2} W_i k(\overline{x}_i, \theta_i, \overline{\mu}) \right)$$

$$\overline{\mu}, \overline{x}_j, \theta_j, \; j = 1, \ldots, ncol + 2 \qquad (8.107)$$

subject to

$$\sum_{i=1}^{ncol+2} A_{ji}\overline{x}_i = f(\overline{x}_j, \theta_j, \overline{\mu}); j = 1, \ldots, ncol + 2 \qquad (8.108)$$

$$h(\overline{x}_j, \theta_j, \overline{\mu}) = 0, \; j = 1, \ldots, ncol + 2 \qquad (8.109)$$

$$g(\overline{x}_j, \theta_j, \overline{\mu}) \leq 0, \; j = 1, \ldots, ncol + 2 \qquad (8.110)$$

$$\overline{x}_0 = \overline{x}_{initial}$$

$$\theta(L) \leq \theta_i \leq \theta(U), \; i = 1, \ldots, ncol + 2$$

$$\overline{\mu}(L) \leq \overline{\mu} \leq \overline{\mu}(U)$$

The above problem involves the solution of a nonlinear optimization problem, which may be achieved using any of the optimization packages, such as GAMS, MINOS, or SQP. The approach is general and can be used even when there are discontinuities in the profiles. However, a note of caution: The discretization increases the dimensionality of the system which, for nonlinear systems, is equivalent to increasing the nonlinearities. This may lead to convergence problems; therefore, good initial values are necessary.

NLP Optimization with Polynomial Approximation for the Control Profiles

Mujtaba and Macchietto (1988) and Farhat et al. (1990) used this approach to solve multifraction maximum distillate problems. The approach involves using polynomial approximation for the control profiles, where the NLP optimizer determines the coefficients of the polynomial. In general, the model for this process is a black box, which provides an implicit relation between the scalar decision variables (coefficients of the polynomial) and the objective function and constraints. This approach is shown in Figure 8.3. Also shown in the figure are the different polynomial approximations used by Farhat to solve the problem of maximizing production of three important fractions, out of a total of five fractions, and the optimal value of the objective function obtained by each approximation. It can be seen that the results of this method depend upon the choice of polynomial.

Combining Maximum Principle and NLP Techniques

The maximum principle formulation of batch distillation and also the dynamic programming and variational calculus, are widely used in the batch distillation literature. However, simultaneous solution of the optimization and optimal control problems is not possible using only the maximum principle. Furthermore, solution of the two-point boundary value problem and the additional adjoint equations with iterative constraint satisfaction can be computationally very expensive, and the nature of the problem does not allow for bounds on the variables. On the other hand, the orthogonal collocation discretization and NLP optimization method can solve the overall optimization problem but involves solution of a higher dimensional system of equations.

A new approach to optimal control problems in batch distillation, proposed in a paper by Diwekar (1992), combines the maximum principle and the NLP techniques. The number of adjoint variables is reduced by introducing quasi-steady state approximations to some of the state variables, thereby reducing the number of adjoint equations. If the number of equality constraints (other than the system model) and the number of reduced adjoint variables are equal, as in the case of batch distillation where one adjoint variable z and one equality constraint are specified in terms of product purity, then neither the Lagrangian formulation of the objective function nor the final boundary conditions of the adjoint variables ($\bar{z}_T = \bar{c}$) is used in the solution. Instead, the final boundary conditions of the adjoint variables are automatically imposed when the equality constraints are satisfied. Maximizing the Hamiltonian provides the functional correlation for the control vectors. In brief, the new algorithm involves solution of the NLP optimization problem for the scalar variables $\bar{u}$ subject to the original model for the state variables 'the adjoint equations' correlation for the control variables, and constraints that implicitly relate to the initial values of the adjoint variables. This algorithm reduces the dimensionality of the problem and avoids the solution of the two-point boundary value problem (Diwekar, 1992). Furthermore, it was shown that for

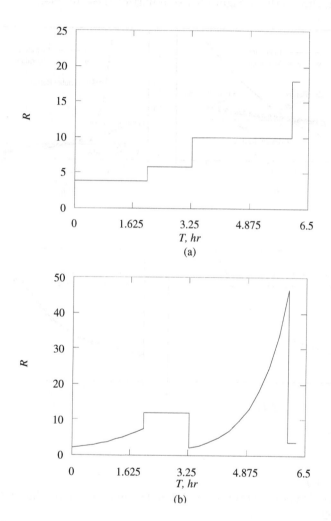

FIGURE 8.3
Optimal Profiles for the Maximum Production Cuts Problem: (a) Optimal Constant Policy, (b) Optimal Exponential/Constant Policy (Reproduced from Farhat et al., 1990)

the batch distillation problem, bounds could be imposed on the control vector by virtue of the nature of the formulation. Diwekar (1992) presented a comparison among the various optimal control problems in batch distillation solved using this method, Figure 8.4 summarizes these results.

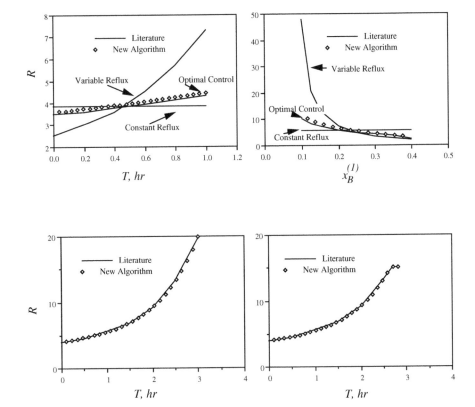

FIGURE 8.4
Solution for Different DAOPs in Batch Distillation (Reproduced from Diwekar, 1992)

The formulation of the DAOP using the new algorithm results in:

$$\text{Optimize} \quad J = j(\overline{x}_T) + \int_0^T k(\overline{x}_t, \theta_t, \overline{\mu}) \, dt \qquad (8.111)$$
$$\overline{\mu}$$

subject to

$$\frac{d\overline{x}_t}{dt} = f(\overline{x}_t, \theta_t, \overline{\mu}) \qquad (8.112)$$

$$\frac{d\bar{z}_t}{dt} = f^*(\bar{x}_t, \theta_t, \bar{\mu}) \tag{8.113}$$

$$\theta_t = H^*(\bar{z}_t, \bar{x}_t) \tag{8.114}$$

$$\bar{z}_0 = h^*(h(\bar{x}_t, \theta_t, \bar{\mu})) \tag{8.115}$$

$$g(\bar{x}_t, \theta_t, \bar{\mu}) \leq 0 \tag{8.116}$$

$$\bar{x}_0 = \bar{x}_{initial}$$

$$\bar{\mu}(L) \leq \bar{\mu} \leq \bar{\mu}(U)$$

$$\theta(L) \leq \theta_t \leq \theta(U)$$

Example 8.5: Formulate the maximum distillate problem presented in Example 8.2 using the combined method and present the complete solution procedure for all the other indices of performance such as the minimum time and maximum profit problems. Present the advantages of the method over that of the other methods.

Solution: The maximum distillate problem in the original form (without considering the Lagrangian formulation as shown in Example 8.2) can be written as:

$$\text{Maximize} \quad -x_T^{(1)} \tag{8.117}$$
$$R_t$$

subject to the following differential equations, and the time-implicit model. The Hamiltonian function, which should be maximized, is:

$$H_t = -z_t^{(1)} \frac{V}{R_t + 1} + z_t^{(2)} \frac{V(x_t^{(2)} - x_D^{(1)})}{(R_t + 1)x_t^{(1)}} \tag{8.118}$$

The adjoint equations are:

$$\frac{dz_t^{(1)}}{dt} = z_t^{(2)} \frac{V(x_t^{(2)} - x_D^{(1)})}{(R_t + 1)(x_t^{(1)})^2}, \quad z_T^{(1)} = -1 \tag{8.119}$$

$$\frac{dz_t^{(2)}}{dt} = -z_t^{(2)} \frac{V\left(1 - \frac{\partial x_D^{(1)}}{\partial x_t^{(2)}}\right)}{(R_t + 1)x_t^{(1)}}, \quad z_T^{(2)} = 0 \tag{8.120}$$

Combining the two adjoint variables $z^{(1)}$ and $z^{(2)}$ into one using $z_t = z_t^{(2)}/z_t^{(1)}$ results in the following adjoint equation:

$$\frac{dz_t}{dt} = -z_t \frac{V\left(1 - \frac{\partial x_D^{(1)}}{\partial x_t^{(1)}}\right)}{(R_t + 1)x_t^{(1)}} - (z_t)^2 \frac{V(x_t^{(2)} - x_D^{(1)})}{(R_t + 1)(x_t^{(1)})^2} \tag{8.121}$$

The optimality condition on the reflux policy $dH_t/dR_t = 0$ leads to

$$R_t = \frac{B_t - z_t(x_B^{(1)} - x_D^{(1)})}{z_t(\partial x_D^{(1)}/\partial R_t)} - 1 \tag{8.122}$$

It should be remembered that this solution (Equation 8.122) is obtained by maximizing the Hamiltonian (maximizing the distillate), which does not incorporate the purity constraint. Hence, use of the final boundary condition ($z_T = 0$) provides the limiting solution resulting in all the reboiler charge instantaneously going to the distillate pot ($R = -\infty$) with the lowest overall purity. Since in this formulation the purity constraint is imposed external to the Hamiltonian, the final boundary condition ($z_T = 0$) is no longer valid.

The solution procedure using the new algorithm is shown in Figures 8.5 and 8.6. The two levels of optimization are: 1) the NLP optimization at the outer loop with respect to the scalar variables μ, and initial value of $R_t(= R_0)^1$, and 2) the inner loop involving calculation of the objective function and the purity constraint for fixed values of the scalar variables and R_0.

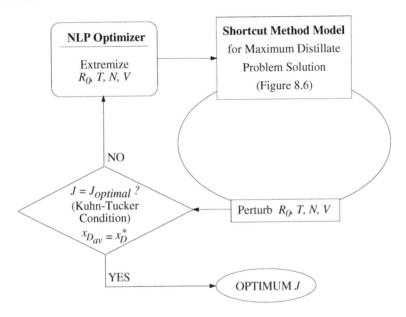

FIGURE 8.5
Combining NLP Optimization and the Maximum Principle (Reproduced from Diwekar, 1992)

Given the scalar variables (e.g., N, V) and the value of R_0, the inner loop initializes the two state variables, and the shortcut method equations allow for the calculation of the other state variables and model parameters (e.g., still and distillate composition). The initial value of the adjoint variable is then calculated from the implicit correlation defining the relationship between the control variable R_t with respect to the adjoint variable z_t and the other model variables. The adjoint equation and state equations are integrated for the next time step and the new value of R_t is calculated. The integration and calculation of the control variable R_t continues until the specified stopping criterion is met. This stopping criterion in Figure 8.6 depends

---

[1] R_0 is used as the decision variable instead of z_0 as proposed in the new algorithm because, with this formulation, it is easier to put bounds on R_0.

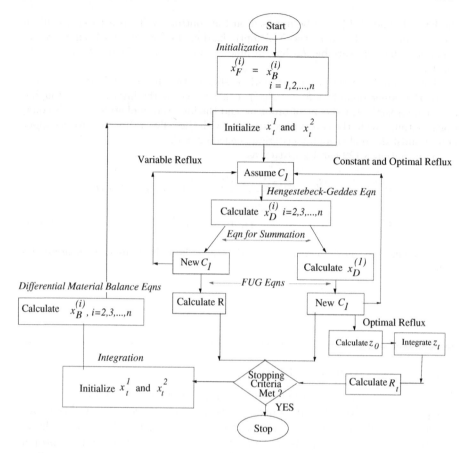

FIGURE 8.6
Unified Shortcut Model for NLP Optimization (Reproduced from Diwekar, 1992)

on the problem at hand. For maximum profit and maximum distillate problems, the final batch time is used as the stopping criterion, and for minimum time problems it is the final amount remaining in the still which marks the end of operation. The values of the objective function and the constraint are calculated at this stage and the control is transferred to the NLP problem, which then computes the new set of scalar variables $\bar{u}$ and R_0.

Since the variable R_0 is independent of the optimal control problem and it has been observed that the following constraint on R_0 is always valid (Converse and Gross, 1963; Keith and Brunet, 1971; Murty et al., 1980; Diwekar et al., 1987; Logsdon et al., 1990; Coward, 1967a; Robinson, 1969; Mayur and Jackson, 1971; Egly et al., 1979; Hansen and Jorgensen, 1986; Kerkhof and Vissers, 1978),

$$R_0 \leq R_t \qquad (8.123)$$

the lower bound $R(L)$ may be imposed on the control profile as a lower bound on the decision variable R_0 in the NLP optimization. In fact, it had been shown in Chapter 5 that the variable R_0 has an inherent lower bound defined by the purity constraint.

The optimal control profile evaluations stop at the stopping criterion. Alternatively, the upper bound to R_t $(= R(U))$ can be used as the intermediate stopping criterion for the optimal control problem, and the integration of state variable equations continues as in the constant reflux case, with the reflux ratio equal to the upper bound, until the real stopping criterion is encountered.

So Equation 8.122 can be written as:

$$R_t = \frac{B_t - z_t(x_B^{(1)} - x_D^{(1)})}{z_t \frac{\partial x_D^{(1)}}{\partial R_t}} - 1 \quad \text{for } R_t \leq R(U)$$

$$R_t = R(U) \qquad \text{for } R_t > R(U) \tag{8.124}$$

This equation allows one to impose the upper bound on the control profile locally.

8.2   Closed-Loop Control

The two traditional batch operation policies, constant reflux and variable reflux policies, involve different control strategies. For the constant reflux policy where the distillate composition is continuously changing, the average distillate composition can only be known at the end of operation unless proper feedback from the operation is obtained. The control of the average distillate composition is, then, of an open-loop control nature. However, the variable reflux policy is inherently a feedback operation because the reflux ratio is constantly adjusted to keep the distillate composition constant. The purpose of designing a closed-loop control scheme is to reduce the sensitivity of the plant to external disturbances. Since batch distillation starts with total reflux to obtain steady state and the distillate is withdrawn after that point, the reflux ratio and distillate composition may oscillate if a controller gain is not properly selected. This is the reason the constant composition control proves to be very challenging. This subsection describes research efforts on closed-loop control problems.

Quintero-Marmol et al. (1991) and Quintero-Marmol and Luyben (1992) proposed and compared several methods for estimating the on-line distillate composition by feedback control under the constant reflux operating policy in a batch rectifier. An extended Luenberger observer for tracking the distillate composition profile was proved to provide the best result.

Bosley and Edgar (1992) considered modeling, control, and optimization aspects of batch rectification using nonlinear model predictive control

(NMPC) and implemented an optimal batch distillation policy that was determined a priori by the off-line optimization. NMPC can determine the set of control moves, which yield the optimal trajectory and allow explicit constraints on inputs, outputs, and plant states. It is known that NMPC is one of the best approaches for distillate composition control; however, the control scheme is computationally intensive because optimization problems are solved inside this control loop. This work was further studied by Finefrock et al. (1994), who studied nonideal binary batch distillation under the variable reflux operating policy. Since the gain space can be changed significantly after a switch to the production phase, they suggested a gain-scheduled PI controller based on NMPC if the instantaneous distillate composition is known. Fileti et al. (2000) also considered NMPC as well as PAD (programmable adaptive controller) and STR (self-turning regulator) for optimization and on-line operation, and suggested that NMPC is generally applicable to any feed mixture conditions.

Besides NMPC, Barolo and Berto (1998) provided a framework for obtaining composition control in batch distillation using a nonlinear internal model control (NIMC) approach. NIMC, proposed by Henson and Seborg (1991), can exactly linearize the system input-output map and be easily tuned by using a single parameter for each component. The distillate composition is estimated by the selected temperature measurements. They also used an extended Luenberger observer for a composition estimator. Although this approach can be reliable and easily implemented, the authors pointed out the problem of selecting the best temperature measurement locations and the problems with using the extended Luenberger observer in the case of a large number of trays of a batch column. For tighter composition control, more research is necessary to develop a robust and fast closed-loop control scheme.

Closed-loop control schemes have also been applied to new column configurations and complex batch systems. Sørensen and Skogestad (1994) presented control schemes for the batch stripper. For the control of the middle vessel column, Barolo et al. (1996) first proposed and examined several control schemes with or without product recycling. They showed the experimental results of the proposed control structures for dual composition control with or without impurity. Farschman and Diwekar (1998) proposed dual composition control in which the two composition control loops can be decoupled if the instantaneous product compositions are known. The degree of interaction between the two composition control loops can be assessed using the relative gain array technique. Phimister and Seider (2000) extended the two composition control loops of continuous distillation to batch distillation having a middle vessel in order to overcome the common problems in the dual composition control of continuous distillation.

Hasebe et al. (1995) proposed a single-loop cascade control system to control the composition of each vessel in the multivessel column. The vessel holdup under total reflux is the manipulated variable, and the reflux flow rate from each vessel is, then, controlled by a simple PI controller. Wittgens et al. (1996) and Skogestad et al. (1997) developed a simple feedback control strategy in which the temperature at the intermediate vessel was controlled by the reflux rates from the vessels, thereby adjusting the vessel holdups indirectly. Further, Furlonge et al. (1999) compared different control schemes, including optimal control problems in terms of energy consumption.

Feedback control in extractive batch distillation has appeared little in literature. Sørensen et al. (1996) proposed a control scheme to track the optimal temperature profile determined a priori by using a conventional PI temperature controller. Monroy-Loperena and Alvarez-Ramirez(2000) extended Barolo and Berto's (Barolo and Berto, 1998) work to reactive batch distillation in which an input-output linearized feedback is proposed.

Future works in closed-loop control problems can involve locating the proper temperature measurement locations, easy parameter tuning, and/or focusing on tracing the optimal profiles, as well as the 'on-spec' products.

References

Akgiray O. and J. Heydeweiller (1990), Numerical methods for optimal control problems,*AIChE Annual Meeting*, Chicago, IL, September.

Aris R. (1961), *The Optimal Design of Chemical Reactors*, Academic Press, London, United Kingdom.

Barolo M. and F. Berto (1998), Composition control in batch distillation: Binary and multicomponent mixtures, *Ind. Eng. Chem. Res.*, **37**, 4689.

Barolo M., F. Berto, S. Rienzi, A. Trotta, and S. Macchietto (1996),Running batch distillation in a column with a middle vessel, *Ind. Eng. Chem. Res.*, **36**, 4612.

Bellman R. (1957), *Dynamic Programming*, Princeton University Press, Princeton, NJ.

Boltyanskii V. G., R. V. Gamkrelidze, and L. S. Pontryagin (1956), On the theory of optimum processes (in Russian), *Doklady Akad. Nauk SSSR*, **110**, no. 1.

Bonny L., P. Floquet, S. Domench and L. Pibouleau (1996), Optimal strategies for batch distillation campaign of different mixtures, *Chemical Engineering and Processing*, **35**, 349.

Bosley J. R. and T. F. Edgar (1992), Application of Nonlinear Model Predictive Control to Optimal Batch Distillation, *IFAC Dynamics and Control of Chemical Reactors (DYCORD+92), Maryland, USA*, 303.

C. Flemming M. and S. B. Jorgensen (1987), Optimal control of binary batch distillation with recycled waste cut, *The Chemical Engineering Journal*, **34**, 57.

Converse A. O. and G. D. Gross (1963), Optimal distillate policy in batch distillation, *Industrial Engineering Chemistry Fundamentals*, **2**, 217.

Coward I. (1967a), The time optimal problems in binary batch distillation, *Chemical Engineering Science*, **22**, 503.

Coward I. (1967b), The time optimal problems in binary batch distillation, a further note, *Chemical Engineering Science*, **22**, 1881.

Cuthrell J. E. and L. T. Biegler (1987), On the optimization of differential-algebraic process systems, *AIChE Journal*, **33**, 1257.

Diwekar U. M. (1992), Unified approach to solving optimal design-control problems in batch distillation, *AIChE Journal*, **38**, 1551.

Diwekar U. M. and K. P. Madhavan (1991), Multicomponent batch distillation column design, *Industrial and Engineering Chemistry Research*, **30**, 713.

Diwekar U. M., R. K. Malik, and K. P. Madhavan (1987), Optimal reflux rate policy determination for multicomponent batch distillation columns, *Computers and Chemical Engineering*, **11**, 629.

Diwekar U. M., K. P. Madhavan, and R. E. Swaney (1989), Optimization of multicomponent batch distillation column, *Industrial and Engineering Chemistry Research*, **28**, 1011.

Egly H., V. Rubby, and V. Seid (1979), Optimum design and operation of batch rectification accompanied by chemical reaction, *Computers and Chemical Engineering*, **3**, 169.

Fan L. T. (1966), *The Continuous Maximum Principle*, John Wiley & Sons, NY.

Farhat S., M. Czernicki, L. Pibouleau, and S. Domench (1990), Optimization of multiple-fraction batch distillation by nonlinear programming, *AIChE Journal*, **36**, 1349.

Farschman C. A. and U. M. Diwekar (1998), Dual composition control in a novel batch distillation column, *Ind. Eng. Chem. Res.*, **37**, 89.

Finefrock Q. B., J. R. Bosley and T. F. Edgar (1994), Gain-scheduled PID control of batch distillation to overcome changing system dynamics, *AIChE Annual Meeting*.

Fileti F. A., S. Cruz and J. Pereira (2000), Control strategies analysis for a batch distillation column with experimental testing, *Chemical Engineering and Processing*, **39**, 121.

Furlonge H. I., C. C. Pantelides, E. Sørensen (1999), Optimal operation of multivessel batch distillation columns, *AIChE J.*, **45**, 781.

Hansen T. T and S. B. Jorgensen (1986), Optimal control of binary batch distillation in tray or packed columns, *Chemical Engineering Journal*, **33**, 151.

Hasebe S., T. Kurooka, and I. Hashimoto (1995), Comparison of the separation performances of a multi-effect batch distillation system and a continuous distillation system, *Preprints of IFAC Symposium on Dynamics and Control of Chemical Reactors, Distillation Columns, and Batch Processes (DYCORD '95)*, Helsingor, Denmark, 249.

Hasebe S. and M. Noda and I. Hashimoto (1999), Optimal operation policy for total reflux and multi-batch distillation systems, *Comp. Chem. Eng.*, *23*, 523.

Henson M. A. and D. E. Seborg (1991), *AIChE J.*,**37**, 1065.

Keith F, and M. Brunet (1971), Optimal operation of a batch packed distillation column, *Canadian Journal of Chemical Engineering*, **49**, 291.

Kerkhof L. H. and H. J. M. Vissers (1978), On the profit of optimum control in batch distillation, *Chemical Engineering Science*, **33**, 961.

Kirk D. E. (1970), *Optimal Control Theory An Introduction*, Prentice-Hall, Englewood Cliffs, NJ.

Li P., G. Wozny, and E. Reuter (1997), Optimization of multiple-fraction batch distillation with detailed dynamic process model,*Institution of Chemical Engineers Symposium Series*,**142**, 289.

Logsdon J. S. and L. T. Biegler (1993), Accurate determination of optimal reflux policies for the maximum distillate problem in batch distillation, *Industrial and Engineering Chemistry Research*, **32**, 692.

Logsdon J. S., U. M. Diwekar, and L. T. Biegler (1990), On the simultaneous optimal design and operation of batch distillation columns, *Transactions of IChemE*, **68**, 434.

Mayur D. N. and R. Jackson (1971), Time optimal problems in batch distillation for multicomponent mixtures columns with holdup, *Chemical Engineering Journal*, **2**,150.

Meski G. and M. Morari (1995),Design and operation of a batch distillation column with a middle vessel, *Comp. Chem. Eng.*, **19**, s597.

Monroy-Loperena R. and J. Alvarez-Ramirez(2000), Output-feedback control of reactive batch distillation column, *Ind. Eng. Chem. Res.*, **39**, 378.

Mujtaba I. M. and S. Macchietto (1988), Optimal control of batch distillation, *IMAC World Conference*, Paris.

Mujtaba I. M. and S. Macchietto (1992), An optimal recycle policy for multicomponent batch distillation, *Computers & Chemical Engineering*, **16**, S273.

Mujtaba I. M. and S. Macchietto (1996),Simultaneous optimization of design and operation of multicomponent batch distillation column–Single and multiple separation duties, *Journal of Process Control*, **6(1)**, 27.

Mujtaba I. M. and S. Macchietto (1997), Efficient optimization of batch distillation with chemical reaction using polynomial curve fitting technique, *Industrial Engineering Chemistry Proceedings Design and Development*, **36**, 2287.

Murty B. S. N., K. Gangiagh, and A. Husain (1980), Performance of various methods in computing optimal control policies, *Chemical Engineering Journal*, **19**, 201.

Phimister J. R. and W. D. Seider (2000), Distillate-bottom control of middle vessel distillation columns, *Ind. Eng. Chem. Res.*, **39**, 1840.

Pontryagin L. S. (1956), Some mathematical problems arising in connection with the theory of automatic control system (in Russian), Session of the Academic Sciences of the USSR on Scientific Problems of Automatic Industry, October 15-20.

Pontryagin L. S. (1957), Basic problems of automatic regulation and control (in Russian), *Izd-vo Akad Nauk SSSR*.

Quintero-Marmol E. and W. L. Luyben (1992), Inferential model-based control of multicomponent batch distillation, *Chem. Eng. Sci.*, **47**, 887.

Quintero-Marmol E., W. L. Luyben and C. Georgakis (1991), Application of an extended Luenberger observer to the control of multicomponent batch distillation, *Ind. Eng. Chem. Res.*, **30(8)**,1870.

Robinson E. R. (1969), The optimization of batch distillation operations, *Chemical Engineering Science*, **24**, 1661.

Robinson E. R. (1970), The optimal control of an industrial batch distillation column, *Chemical Engineering Science*, **25**, 921.

Skogestad S., B. Wittgens, R. Litto, and E. Sørensen (1997), Multivessel batch distillation, *AIChE J.*, **43**, 971.

Sørensen E. and S. Skogestad (1994), *Comp. Chem. Eng.*, **18 Suppl.**, s391.

Sørensen E., S. Macchietto, G. Stuart, and S. Skogestad (1996), Optimal control and on-line operation of reactive batch distillation, *Comp. Chem. Eng.*, **20**, 1491.

Wajge R. M. and G. V. Reklaitis(1998), An optimal campaign structure for multicomponent batch distillation with reversible reaction, *Ind. Eng. Chem. Res.*, **37**, 1910.

Wittgens B., R. Litto, E. Sørensen, and S. Skogestad (1996), Total reflux operation of multivessel batch distillation, *Comp. Chem. Eng.*, **20**, s1041.

Exercises

8.1 A performance equation for a simple process is given by:

$$\frac{dx_1}{dt} = -ax_1 + \theta_t, x_1(o) = \alpha, 0 \leq t \leq T$$

The objective is to maximize the following index of performance.

$$J = \frac{1}{s}\int_0^T \left[(x_1)^2 + (\theta_t)^2\right] dt$$

Solve the above problem using the calculus of variations.

8.2 Solve the problem given in Problem 8.1 using the dynamic programming method and compare the results.

8.3 Solve the problem given in Problem 8.1 using the maximum principle method and compare the results with Problems 8.1 and 8.2.

8.4 Formulate the minimum time problem using all three methods.

8.5 Derive Equations 8.30 to 8.33 for the maximum distillate problem from the Euler–Lagrangian formulation.

8.6 Formulate the maximum production cut problem given by Farhat et al. (1990), using the dynamic programming method.

8.7 Discuss the advantages and disadvantages of the different methods presented in this chapter.

9

CONSIDERATION OF UNCERTAINTY

CONTENTS

One of the advantages of batch distillation is its flexibility to deal with changing feed composition and product specification. However, this flexibility poses the problem of uncertainties and variabilities in these quantities. Further, batch distillation is used for pharmaceutical industries where new drugs and chemicals are invented regularly. The thermodynamic properties of these chemicals are not known precisely. This results in thermodynamic uncertainties (Ulas and Diwekar, 2004)). Moreover, there are measurement uncertainties and disturbances in the process.

Rico-Ramirez et al. (2003) describes three kinds of uncertainties encountered in batch processes. These are given below.

1. Uncertainty with respect to the model parameters: These parameters are a part of the deterministic model and not actually subject to randomness. Theoretically, their value is an exact number. The uncertainty results from the impossibility of modeling the physical behavior of the system exactly. For example, thermodynamic uncertainties.

2. Uncertainty in the input variables: This kind of uncertainty originates from the random nature and unpredictability of certain process inputs (e.g., feed composition uncertainty).

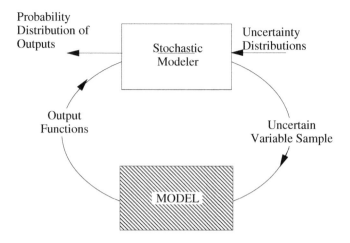

FIGURE 9.1

The Stochastic Modeling Framework

> 3. Uncertainty in the initial conditions: This kind of uncertainty creates problem in the dynamics of the column, e.g., initial total reflux condition uncertainty (Barolo and Cengio, 2001).

There are instances of all three sources of uncertainties in batch distillation. Due to the unsteady state nature of the process, these uncertainties can result in dynamic uncertainties (Ulas and Diwekar, 2004). Therefore, uncertainty analysis is a crucial step in batch distillation design, optimization, and control and is the focus of this chapter.

9.1 Static Uncertainties and Stochastic Modeling

Static uncertainties like feed composition variability, thermodynamic parameter uncertainty, and measurement uncertainty are normally represented by probability distributions. Inclusion of uncertainties in a deterministic model results in a stochastic model. Stochastic modeling is an iterative procedure that consists of the following four steps (Diwekar and Rubin, 1991), as shown in Figure 9.1.

> 1. Uncertainty quantification that involves specifying uncertainties in key input parameters in terms of probability distributions.
>
> 2. Sampling distribution of the specified parameter in an iterative fashion.
>
> 3. Propagating the effects of uncertainties through the model.

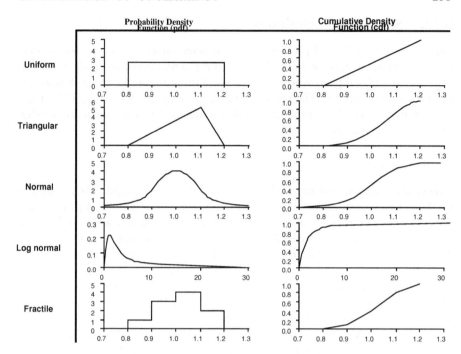

FIGURE 9.2
Examples of Probabilistic Distribution Functions for Stochastic Modeling

4. Applying statistical techniques to analyze the outputs.

9.1.1 Uncertainty Quantification

To accomodate the diverse nature of uncertainty, different distributions can be used. The type of distribution for an uncertain variable is a function of the amount of data available and the characteristic of the distribution function. The simplest distribution for an uncertain variable is a uniform distribution, which has a constant probability. This means that the uncertain variable can take any value within an interval [a,b] with equal probability. On the other hand, if the uncertain variable is represented by a normal (Gaussian) distribution, there is a symmetric but equal probability that the value of the uncertain variable will be above or below a mean value. In log-normal or some triangular distributions, there is a higher probability that the value of an uncertain variable will be on one side of the median, resulting in a skewed shape. A beta distribution provides a wide range of shapes and is a very flexible means of representing variability over a fixed range. In some special cases, user-supplied distributions are used, e.g., chance distribution. Different examples of probability distributions are given in Figure 9.2.

9.1.2   Sampling Techniques

Sampling is a statistical procedure that involves selecting a limited number of observations, states, or individuals from a population of interest. A sample is assumed to be representative of the whole population to which it belongs. Instead of evaluating all the members of the population, which would be time consuming and costly, sampling techniques are used to infer some knowledge about the population.

Monte Carlo Sampling

One of the simplest and most widely used methods for sampling is the Monte Carlo method. Monte Carlo methods are numerical methods that provide approximate solutions to a variety of physical and mathematical problems by random sampling. The name Monte Carlo, which was suggested by Nicholas Metropolis, takes its name from a city in the Monaco principality, which is famous for its casinos, because of the similarity between statistical experiments and the random nature of games of chance, e.g., roulette. Monte Carlo methods were originally developed for the Manhattan Project during World War II to simulate probabilistic problems related to random neutron diffusion in fissile material. Although they were limited by the computational tools of that time, they became widely used in many branches of science after electronic computers were built in 1945. The first publication that presents the Monte Carlo algorithm is probably by Metropolis and Ulam (1949).

One remarkable feature of this sampling technique is that the error bound for output estimation is not dependent on the dimension k. However, this bound is probabilistic, which means that there is never any guarantee that the expected accuracy will be achieved in a concrete calculation. The success of a Monte Carlo calculation depends on the choice of an appropriate random sample. The required random numbers and vectors are generated by the computer in a deterministic algorithm. Therefore, these numbers are called pseudo-random numbers or pseudo-random vectors.

Latin Hypercube Sampling

Stratification is the grouping of the members of a population into equal or unequal probability areas (strata) before sampling. The strata must be mutually exclusive, which means that every element in the population must be assigned to only one stratum. Also, no population element is excluded. It is required that the proportion of each stratum in the sample should be the same as in the population. Latin hypercube sampling is one form of stratified sampling that can yield more precise estimates of the distribution function (McKay et al., 1979) and therefore reduce the number of samples required to improve computational efficiency. In LHS, the range of each uncertain parameter x_i is subdivided into non-overlapping intervals of equal probability. One value from each interval is selected at random with respect to the probability distribution in the interval. The n values thus obtained for x_1 are paired in a random manner (i.e., equally likely combinations) with n values of x_2. These n values are then combined with n values of x_3 to form n triplets, etc., until n k-tuplets are formed.

In median Latin hypercube sampling (MLHS), which is a variant of LHS, the midpoint of the intervals is chosen to sample the uncertain variables. The main drawback of this stratification scheme in LHS and MLHS is that it is uniform in one dimension (1D) and does not provide uniformity properties in k dimensions. Sampling based on quadrature, cubature techniques, or collocation techniques face similar drawbacks.

Hammersley Sequence Sampling

Hammersley sequence sampling is an efficient sampling technique developed by Diwekar and co-workers (Diwekar and Kalagnanam, 1997; Diwekar, 2008) based on quasi-random numbers. The Hammersley sequence sampling uses Hammersley points to uniformly sample a unit hypercube and inverts these points over the joint cumulative probability distribution to provide a sample set for the variables of interest. As shown in Figure 9.3, Hammersley sequence sampling technique uses an optimal design scheme for placing n points on a k-dimensional hypercube. This scheme ensures that it is more representative of the population showing uniformity properties in multidimensions, unlike Monte Carlo, Latin hypercube, and its variant MLHS techniques. A qualitative picture of the uniformity properties of the different sampling techniques on a unit square is presented in Figure 9.3. It is clearly observed that HSS shows better uniformity than other stratified sampling techniques, e.g., LHS, which are uniform along a single dimension only and do not guarantee a homogeneous distribution of points over the multivariate probability space. One of the main advantages of Monte Carlo methods is that the number of samples required to obtain a given accuracy of estimates does not scale exponentially with the number of uncertain variables. HSS preserves this property of Monte Carlo. HSS and its variants have been found to be 3 to 100 times faster than traditional MCS or LHS techniques and is a preferred technique for stochastic modeling or stochastic optimization. However, many packages like MATLAB® and Crystall Ball do not have this technique embedded in them. These packages have MCS or LHS techniques available.

9.1.3   Output Analysis

In general, the output parameters after propagating the samples through the model are analyzed in terms of a cumulative distribution functional such as the expected value, median, mode, variance, or fractiles. Figures 9.4 and 9.5 show the expected value or mean, mode, median, variance, and fractiles (at various probabilities) for a particular output parameter whose probability density

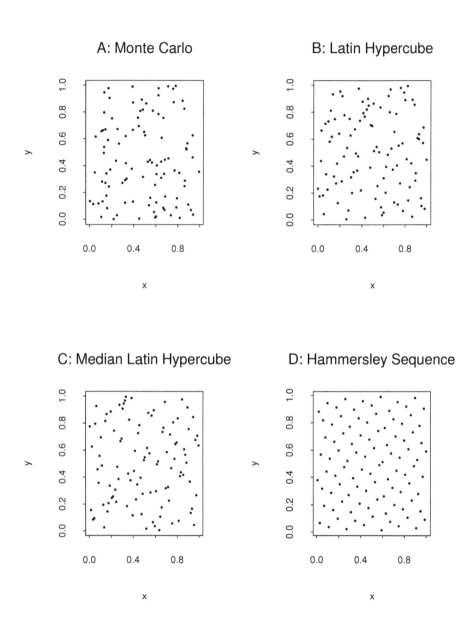

FIGURE 9.3
Comparison of Sample Points (100) on a Unit Square Using Various Sampling
Techniques for Two Uniform Distributions

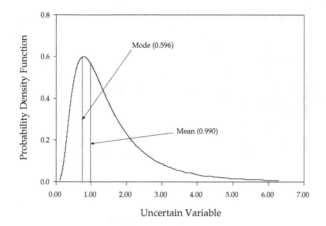

FIGURE 9.4
Different Probabilistic Performance Measures (PDF)

function is shown in Figure 9.4 and cumulative density function (CDF) in Figure 9.5.

Most of the problems related to uncertainties in batch distillation are either optimization or optimal control problems under uncertainty where stochastic modeling steps are used for uncertainty analysis. However, these problems are more difficult than just stochastic modeling and are described next. The description is derived from my book on optimization (Diwekar, 2008).

9.2   Optimization under Uncertainty

The literature on optimization under uncertainties very often divides the problems into categories such as "wait and see", "here and now", and "chance constrained optimization" (Vajda, 1972; Nemhauser et al., 1989). In wait and see we wait until an observation is made on the random elements, and then solve the deterministic problem. This is similar to the wait and see problem of Madansky (1960), originally called "Stochastic Programming" by Tintner (1955), and is not in a sense one of decision analysis. In decision making, the decisions have to be made here and now about the activity levels. The here and now problem involves optimization over some probabilistic measure. By this definition, chance constrained optimization problems can be included in this particular category of optimization under uncertainty. Chance constrained optimization involves constraints that are not expected to be always satisfied,

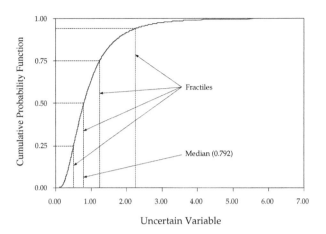

FIGURE 9.5
Different Probabilistic Performance Measures (CDF)

only in a proportion of cases, or with given probabilities. These various categories require different methods for obtaining their solutions.

9.2.1 Here and Now Problems

Stochastic optimization gives us the ability to optimize systems in the face of uncertainties. The here and now problems require that the objective function and constraints be expressed in terms of some probabilistic representation (e.g., expected value, variance, fractiles, most likely values). For example, in chance constrained programming, the objective function is expressed in terms of expected value, and the constraints are expressed in terms of fractiles (probability of constraint violation). In Taguchi's offline quality control method (Taguchi, 1986; Diwekar and Rubin, 1991), the objective is based on variance. These problems can be classified as here and now problems.

The here and now problem, where the decision variables and uncertain parameters are separated, can then be viewed as

$$\text{Optimize}_{x} \quad J \;=\; P_1(j(x,u)) \tag{9.1}$$

subject to

$$P_2(h(x,u)) \;=\; 0 \tag{9.2}$$

$$P_3(g(x,u) \geq 0) \;\geq\; \alpha \tag{9.3}$$

where u is the vector of uncertain parameters and P represents the cumulative distribution functional such as the expected value, mode, variance, or fractiles.

Unlike the deterministic optimization problem, in stochastic optimization one has to consider the probabilistic functional of the objective function and constraints. The generalized treatment of such problems is to use probabilistic or stochastic models described earlier instead of the deterministic model inside the optimization loop.

Figure 9.6a represents the generalized solution procedure, where the deterministic model is replaced by an iterative stochastic model with a sampling loop representing the discretized uncertainty space. The uncertainty space is represented in terms of the moments such as the mean or the standard deviation of the output over the sample space of N_{samp}, as given by the following equations (Equations (9.4) and (9.5)).

$$E(z(x,u)) \; = \; \sum_{k=1}^{N_{samp}} \frac{z(x,u_k)}{N_{samp}} \tag{9.4}$$

$$\sigma^2(z(x,u)) \; = \; \sum_{k=1}^{N_{samp}} \frac{(z(x,u_k) \; - \; \bar{z})^2}{N_{samp}} \tag{9.5}$$

where $\bar{z}$ is the average value of z. E is the expected value and σ^2 is the variance.

In chance constrained formulations, the uncertainty surface is translated into input moments, resulting in an equivalent deterministic optimization problem. On the other hand, the worst-case robust optimization formulation leads to a bi-level min-max optimization problem. These two formulations can be considered as special cases of here and now problems.

Chance Constrained Programming

In the chance constrained programing (CCP) method, some of the constraints likely need not hold. Chance constrained problems can be represented as follows.

$$\underset{x}{\text{Optimize}} \quad J \; = \; P_1(j(x,u)) \; = \; E(z(x,u)) \tag{9.6}$$

subject to
$$P(g(x) \leq u) \; \leq \; \alpha \tag{9.7}$$

In the above formulation, Equation (9.7) is the chance constraint. In the chance constraint formulation, this constraint (or constraints) is (are) converted into a deterministic equivalent under the assumption that the distribution of the uncertain variables u is a stable distribution. Stable distributions

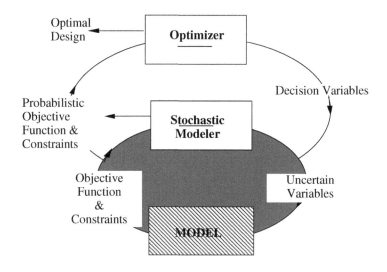

(a) Here and n ow

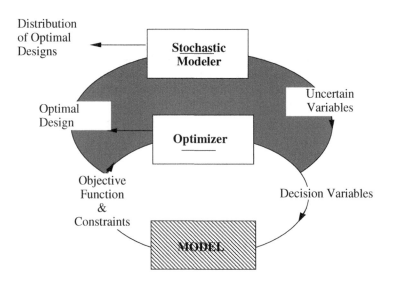

(b) Wait and s ee

FIGURE 9.6
Pictorial Representation of the Stochastic Programming Framework

are such that the convolution of two distribution functions $F(x - m_1/v_1)$ and $F(x - m_2/v_2)$ is of the form $F(x - dmu/v)$, where m_i and v_i are two parameters of the distribution (Luckacs, 1970). Normal, Cauchy, uniform, and chi-square are all stable distributions that allow the conversion of probabilistic constraints into deterministic ones. The deterministic constraints are in terms of moments of the uncertain variable u (input uncertainties). For example, if constraint g in Equation (9.7) has a cumulative probability distribution F then the deterministic equivalent of this constraint is given below.

The deterministic equivalent of the chance constraint Equation 9.7 is:

$$g(x) \leq F^{-1}(\alpha) \tag{9.8}$$

where F^{-1} is the inverse of the cumulative distribution function F.

The major restrictions in applying the CCP formulation include that the uncertainty distributions should be stable distribution functions, the uncertain variables should appear in the linear terms in the chance constraint, and that the problem needs to satisfy the general convexity conditions. The advantage of the method is that one can apply deterministic optimization techniques to solve the problem.

The Worst-case Robust Optimization

A robust optimal design is one with the best worst-case performance. If the optimization under uncertainty problem involves the following optimization problem given by Equations 9.10-9.12, then the worst-case robust optimization formulation results in the min-max problem given by Equations 9.13-9.15.

$$\text{Min.} \quad J = j(x, u, \theta) \tag{9.9}$$
$$x \tag{9.10}$$

subject to

$$h(x, u, \theta) = 0 \tag{9.11}$$

$$g(x, u, \theta) \geq 0 \tag{9.12}$$

$$\text{Min} \quad J = \text{Max}(j(x, u, \theta))$$
$$x, u \qquad\qquad \theta \tag{9.13}$$

$$\text{Max}(g(x, u, \theta)) \leq 0$$
$$\theta \tag{9.14}$$

subject to

$$h(x, u, \theta) = 0 \tag{9.15}$$

9.2.2 Wait and See

In contrast to here and now problems, which yield optimal solutions that achieve a given level of confidence, wait and see problems involve a category of formulations that shows the effect of uncertainty on optimum design. A wait and see problem involves deterministic optimal decisions at each scenario or random sample, equivalent to solving several deterministic optimization problems. The generalized representation of this problem is given below.

$$\text{Optimize} \quad Z = z(x, u*) \qquad (9.16)$$
$$x$$

subject to

$$h(x, u*) = 0 \qquad (9.17)$$

$$g(x, u*) \leq 0 \qquad (9.18)$$

where $u*$ is the vector of values of uncertain variables corresponding to each scenario or sample.

This optimization procedure is repeated for each sample of uncertain variables u and a probabilistic representation of the outcome is obtained.

Figure 9.6b represents the generalized solution procedure, where the deterministic problem forms the inner loop, and the stochastic modeling forms the outer loop. The measurement based optimization problems in batch processing reported by Srinivasan et al. (2002) come under this category.

From Figures 9.6 it is clear that by simply interchanging the position of the uncertainty analysis framework and the optimization framework, one can solve many problems in the stochastic optimization and stochastic programming domain (Diwekar, 2008).

9.3 Uncertainty in Batch Distillation

As stated earlier, the three types of uncertainties in batch processing are all present in batch distillation. The first problem dealing with uncertainties in batch distillation was presented by Diwekar and Kalagnanam in 1997 (Diwekar and Kalagnanam, 1997) where they considered uncertainties in thermodynamic parameter, relative volatility (α), and input parameters feed composition, reflux ratio, vapor rate, and amount of feed. These uncertainties were modeled as a normal distribution. The problem that was solved is the problem of quality control. This problem involved considering objective function based on variance and solving the here and now problem. Figure 9.7 shows how the objective function based on variance is optimized in this problem. This problem is also called an off-line quality control or parameter design problem (Diwekar and Rubin, 1994). The HSS method was used to solve the problem.

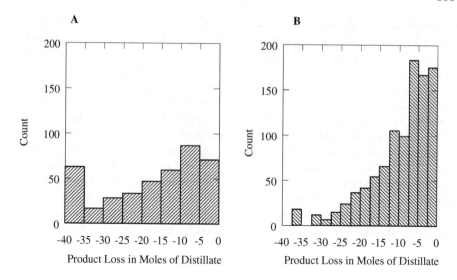

FIGURE 9.7

Reducing Product Loss Based on Variance (Diwekar and Kalagnanam, 1997)

Bernard and Saraiva (1998) extended this problem and included both parameter and tolerance design. They considered uncertainties in relative volatility (uniform distribution), reflux ratio (normal distribution), total feed (uniform distribution), and feed composition (uniform distribution).

Arellano-Garcia et al. (2003) used the chance constrained formulation for batch distillation. They extended the steady state chance constraint formulation presented in an earlier section to a dynamic system in which some dependent variables at certain times are to be constrained by predefined probability. The main idea of this method is the employment of the monotone relation between output constraints and uncertain variables, so that the probabilities and their gradients can be achieved by numerical integration of the probability density function of the multivariate uncertain variables by collocation on finite elements. The approach involved new efficient algorithms for realizing the required reverse projection and hence the probability and gradient computation with an optimal number of collocation points so that the original idea is now applicable to dynamic optimization problems with a larger scale. In this problem, uncertainties in kinetic parameters and efficiency are considered for a reactive batch distillation column. Gaussian or normal distributions are used for uncertainties. A similar approach is used by Kooks (2003) but instead of using the collocation method, the Monte Carlo method is used for integration similar, to the here and now problems. In this paper, uniform distributions were used for uncertainties in relative volatility, feed composition, reflux ratio, and vapor rate.

Diehl et al. (2006) and Diehl et al. (2008) presented the worst-case robust optimization formulation where they used uncertainty in relative volatility, α, and feed composition. In all these studies ideal thermodynamics are used to represent the vapor liquid equilibria. However, many real world systems shows nonideality (e.g., azeotropic systems) in their thermodynamic behavior. The nonideality in the liquid side of the vapor liquid equilibria can be captured using a multiplicative parameter called the activity coefficient and predicted by methods like UNIFAC and WILSON. Ulas and Diwekar (2004, 2006) and Ulas et al. (2005) considered uncertainties in the activity coefficients for nonideal systems and used a stochastic optimal control formulation. Figure 9.8 shows the uncertainty quantification for the activity coefficients for the ethanol-water system reported by Ulas and Diwekar (2004). The figure shows distributions of the uncertainty factor (UF) defined as the ratio of the activity coefficient obtained using the thermodynamic model to the experimental activity coefficient for ethanol (top figure) and for water (bottom figure).

9.3.1   Dynamic Uncertainties

The unsteady state nature of batch distillation converts static uncertainties into dynamic uncertainties. For example, consider the case of emerging processes for bio-diesel. Rape methyl ester (RME) is a form of bio-fuel which is regarded as an alternative to petroleum based diesel. For the production of this bio-fuel, methanol is used as a solvent and it is recovered by distillation at the end of the process. Figure 9.9 shows the results of rigorous model simulation for the separation of methanol from bio-diesel in a batch column (Ulas and Diwekar, 2004). Since no data for the vapor-liquid equilibrium is available for the separation of RME from methanol, the individual components or the fatty acids which constitute the rape methyl ester are lumped to predict the phase equilibrium in this simulation. This figure demonstrates the uncertainties involved in predicting relative volatility of this mixture. First, note that there is a significant amount of change in relative volatility for this mixture and second, the change is time dependent because of batch processing.

As stated earlier, it is a common practice to use probability distribution functions like normal, log-normal, uniform, to model uncertainties. However, these distributions are used for scalar parameter uncertainties. Modeling dynamic or time-dependent uncertainties is a difficult task. Recently, the author (Diwekar 2003, 2008) presented basic concepts for modeling time-dependent uncertainties. These concepts are derived from the financial and economics literature where time-dependent uncertainties dominate. The following paragraphs present these concepts. This is followed by a subsection that uses an analogy between stock prices and the relative volatility in batch distillation to illustrate the usefulness of modeling time-dependent uncertainties using these concepts.

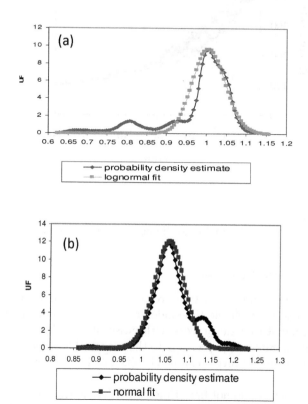

FIGURE 9.8

Probability Distribution Functions for Activity Coefficient Uncertainty in (a) Ethanol and (b) Water (Ulas and Diwekar, 2004)

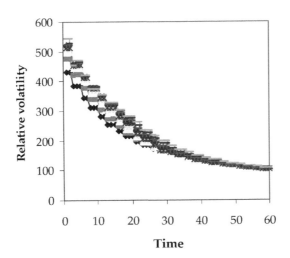

FIGURE 9.9
Uncertainties in Relative Volatility for Separation of Methanol from Bio-Diesel

9.3.1.1 Stochastic Processes

A stochastic process is a variable that evolves over time in an uncertain way. A stochastic process in which the time index t is a continuous variable is called a continuous-time stochastic process. Otherwise, it is called a discrete-time stochastic process. Similarly, according to the conceivable values for x_t (called the states), a stochastic process can be classified as being continuous-state or discrete-state.

Stochastic processes do not have time derivatives in the conventional sense and, as a result, they cannot always be manipulated using the ordinary rules of calculus. This is because, in general, the solution to a stochastic differential equation is not a single value for the function, but rather a probability distribution. As a result, the typical mathematical techniques used to solve the dynamics cannot be directly applied. To work with stochastic processes, one must make use of Ito's lemma. This lemma, called the fundamental theorem of stochastic calculus, allows us to differentiate and to integrate functions of stochastic processes named Ito processes.

One of the simplest examples of a stochastic process is the random walk process. The Wiener process, also called a Brownian motion, is a continuous limit of the random walk and is a continuous-time stochastic process. The Wiener process can be used as a building block to model an extremely broad range of variables that vary continuously and stochastically through time.

For example, consider the price of a technology stock. It fluctuates randomly, but over a long time period has had a positive expected rate of growth that compensated investors for risk in holding the stock. Can the stock price be represented as a Wiener process? The following paragraphs establish that stock prices can be represented as a Wiener process, as it has these important properties.

1. It satisfies the Markov property. The probability distribution for all future values of the process depends only on its current value. Stock prices can be modeled as Markov processes, on the grounds that public information is quickly incorporated in the current price of the stock and past patterns have no forecasting value.

2. It has independent increments. The probability distribution for the change in the process over any time interval is independent of any other time interval (nonoverlapping).

3. Changes in the process over any finite interval of time are normally distributed, with a variance that is linearly dependent on the length of time interval dt.

From the example of the technology stock above, it is easier to show that the variance of the change distribution can increase linearly. However, given that stock prices can never fall below zero, price changes cannot be represented as a normal distribution. However, it is reasonable to assume that changes in the logarithm of prices are normally distributed. Thus, stock prices can be represented by the logarithm of a Wiener process.

As stated earlier, stochastic processes do not have time derivatives in the conventional sense and, as a result, they cannot be manipulated using the ordinary rules of calculus as needed to solve stochastic optimal control problems. Ito provided a way around this by defining a particular kind of uncertainty representation based on the Wiener process.

An Ito process is a stochastic process $x(t)$ on which its increment dx is represented by the equation:

$$dx = a(x,t)dt + b(x,t)dz \qquad (9.19)$$

where dz is the increment of a Wiener process, and $a(x,t)$ and $b(x,t)$ are known functions. By definition, $E[(dz)] = 0$ and $(dz)^2 = dt$ where E is the expectation operator and $E[dz]$ is interpreted as the expected value of dz.

The simplest generalization of Equation (9.19) is the equation for Brownian motion with drift given by

$$dx = \alpha dt + \sigma dz \qquad \text{Brownian motion with drift} \qquad (9.20)$$

where α is called the drift parameter, and σ is the variance parameter. The discretized form of Equation (9.20) is the following:

$$x_t = x_{t-1} + \alpha \Delta t + \sigma \epsilon_t \sqrt{\Delta t} \qquad (9.21)$$

where ϵ_t is normally distributed with a mean of 0 and a standard deviation of 1.0. Figure 9.10 shows the sample paths of Equation (9.20). Over any time interval Δt, the change in x, denoted by Δx, is normally distributed and has an expected value variance:

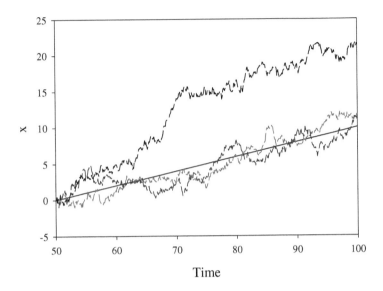

FIGURE 9.10
Sample Paths for a Brownian Motion with Drift

$$E[\Delta t] \quad = \quad \alpha \Delta t \qquad (9.22)$$
$$\nu[\Delta t] \quad = \quad \sigma^2 \Delta t \qquad (9.23)$$

For calculation of α, the average value of the differences in x ($E[x_t - x_{t-1}]$) is computed. Then this value is divided by the time interval Δt to obtain α. On the other hand, for σ, the variance of the differences in x is found and divided by the time interval Δt. Then the square root of this value is computed.

Other examples of Ito processes are geometric Brownian motion with drift (Equation (9.24) given below) and the mean reverting process (Equation (9.28), Figure 9.11).

$$dx \quad = \quad \alpha x dt \, + \, \sigma x dz \quad \text{geometric Brownian motion with drift} \qquad (9.24)$$

In geometric Brownian motion, the percentage changes in x and $\Delta x/x$ are normally distributed (absolute changes are log-normally distributed). We can write Equation (9.24) in the following form if we write $F(x) = \log x$.

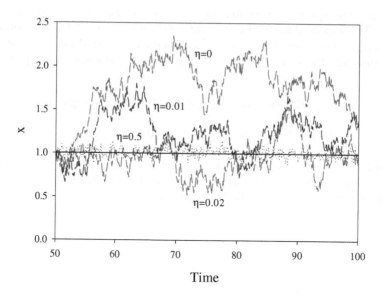

FIGURE 9.11
Sample Paths of a Mean Reverting Process

$$dF = (\alpha - \frac{\sigma^2}{2})dt + \sigma dz \qquad (9.25)$$

Over the time interval t, the change in the logarithm of x is normally distributed with mean $(\alpha - (\sigma^2/2))t$ and variance $\sigma^2 t$. We can estimate the parameters of this Ito process following this procedure. First we find the variance of the changes in the logarithm of x, $(\ln x_t - \ln x_{t-1})$. When we divide this value by Δt, we can obtain σ^2. Once we know the value of σ, we can then calculate the mean value of the changes in the logarithm of x, $(\ln x_t - \ln x_{t-1})$, which is equal to $(\alpha - (\sigma^2/2))t$. From that value, we can calculate α. It was shown that for the absolute value of x, Equations (9.26) (expected value) and (9.27) (variance) hold true:

$$
\begin{aligned}
E[x(t)] &= x_0 . \exp(\alpha t) & (9.26)\\
\nu[x(t)] &= x_0{}^2 . \exp(2\alpha t(\exp \sigma^2 t - 1)) & (9.27)
\end{aligned}
$$

In mean reverting processes, the variable may fluctuate randomly in the short run, but in the longer run it will be drawn back toward the marginal value of the variable:

$$dx = \eta(x_{avg} - x)dt + \sigma dz \quad \text{mean reverting process} \qquad (9.28)$$

where η is the speed of reversion and x_{avg} is the nominal level to which x reverts. The expected value of change in x depends on the difference between x and x_{avg}. If the current value of x is x_0, then the expected value of x at any future time and the variance of $x_t - x_{avg}$ is given by the following equations.

$$E[x(t)] = x_{avg} + (x_0 - x_{avg}) \exp(-\eta t) \qquad (9.29)$$

$$\nu[x_t - x_{avg}] = \frac{\sigma^2}{2\eta}(1 - \exp(-2\eta t)) \qquad (9.30)$$

From these equations it can be observed that the expected value of x_t converges to x_{avg} as t becomes large and the variance converges to $\sigma^2/2\eta$. We can write Equation (9.28) in the following form.

$$x_t - x_{t-1} = \eta x_{avg} \Delta t - \eta x_{t-1} \Delta t + \sigma \epsilon_t \sqrt{\Delta t} \qquad (9.31)$$

$$x_t - x_{t-1} = C_1 + C_2 x_{t-1} + e_t \qquad (9.32)$$

In order to estimate the parameters we can run the regression with the available discrete time data (Equation (9.32)). In this equation, $C_1 = \eta x_{avg} \Delta t$, $C_2 = -\eta \Delta t$, and $e_t = \sigma \epsilon_t \sqrt{\Delta t}$. From the standard error of regression e_t, one can calculate the standard deviation σ.

In Equation (9.28), if the variance rate grows with x, we obtain the geometric mean reverting process:

$$dx = \eta(x_{avg} - x)dt + \sigma x dz \quad \text{geometric mean reverting process} \quad (9.33)$$

The procedure for parameter estimation for this process is the following. We can write Equation (9.33) in the following form.

$$x_t - x_{t-1} = \eta x_{avg} \Delta t - \eta x_{t-1} \Delta t + \sigma x_{t-1} \epsilon_t \sqrt{\Delta t} \qquad (9.34)$$

If we divide both sides by x_{t-1}, Equation (9.35) is obtained:

$$\frac{x_t - x_{t-1}}{x_{t-1}} = \frac{C_1}{x_{t-1}} + C_2 + e_t \qquad (9.35)$$

In this equation, $C_1 = \eta x_{avg} \Delta t$, $C_2 = -\eta \Delta t$, and $e_t = \sigma \epsilon_t \sqrt{\Delta t}$. By running this regression using the available discrete time data we can find the values of C_1 and C_2, which enable us to predict the parameters in Equation (9.33). Again, from the standard error of regression, one can calculate the standard deviation σ.

9.3.2   Relative Volatility: An Ito Process

What is common between the technology stock price example given earlier and the uncertainty in the relative volatility parameter in the batch distillation models?

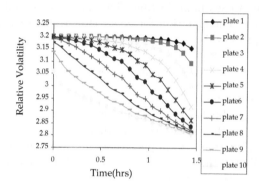

FIGURE 9.12
Relative Volatility, Ideal System as a Function of Time and Number of Plate
(Rico-Ramirez et al., 2003)

1. Both have time-dependent variations. The technology stock fluctuates around the mean randomly, but over time has a positive
   expected rate of growth. Relative volatility for an ideal system, on
   the other hand, fluctuates around the geometric mean across the
   column height, but over a time period the mean decreases (Figure
   9.12 shows the relative volatility fluctuations for a pentane-hexane
   system).

2. Similar to stock prices, relative volatility can be modeled as a
   Markov process because, at any time period, the value of relative
   volatility depends only on the previous value. The changes for both
   are nonoverlapping.

3. Whether uncertainty in the relative volatility parameter can be represented by a Wiener process can be shown with some simple numerical experiments where the data are generated from a rigorous

simulation model (proxy for experiments) for various thermodynamic systems.

In this section, I present the result of a simple numerical experiment to show that the behavior of the relative volatility in a batch column can indeed be represented as an Ito process. Here I take two examples; the first one is the relative volatility of an ideal system with the pentane-hexane mixture (Rico-Ramirez et al., 2003). Figure 9.12 shows the behavior of the relative volatility with respect to time and plate number for this example. A rigorous simulation with the simulation package $MultiBatchDS^{TM}$(Diwekar, 1996) was performed in a batch column to obtain the behavior of the relative volatility with respect to time. As we know, the relative volatility is different for each plate of the column at each point in time. This can be captured by a geometric Brownian motion. The equation for geometric Brownian motion (special instance of an Ito process) is

$$d\alpha = \alpha\beta dt + \alpha\sigma dz \qquad (9.36)$$

where β and σ are constants.

Equation (9.36) establishes that the changes in relative volatility are log-normally distributed with respect to time. In fact, by using Ito's lemma, it can be shown that Equation 9.36 implies that the change in the logarithm of α is normally distributed (for a finite time interval t) with mean $(\beta - 1/2\sigma^2)\,t$ and variance $\sigma^2 t$, resulting in Equation (9.37).

$$d(\ln \alpha) = \left(\beta - \frac{1}{2}\sigma^2\right) dt + \sigma dz \qquad (9.37)$$

In Equations (9.36) and (9.37), dz is defined as

$$dz = \epsilon_t \sqrt{dt}$$

where ϵ_t is drawn from a normal distribution with a mean of zero and unit standard deviation. By using the time series data for relative volatility, the natural logarithm of relative volatility for a fixed time interval can be used to obtain the mean and variance of the underlying normal distribution. It has been found that the data shown in Figure 9.12 fit well with this representation. Then, $\alpha(t)$ can be calculated by using Equation (9.38) given below.

$$\alpha_t = (1 + \beta\,\Delta t)\,\alpha_{t-1} + \sigma\,\alpha_{t-1}\,\epsilon_t\,\sqrt{\Delta t} \qquad (9.38)$$

Consider now a system of a nonideal mixture, such as ethanol-water studied by Ulas and Diwekar (2004). This mixture results in a different relative volatility profile from an ideal mixture, as shown in Figure 9.13a. It was found that this behavior can be best modeled with a geometric mean reverting process rather than a geometric Brownian motion (Figure 9.13b). The equation for the geometric mean reverting process is:

$$d\alpha = \eta(\alpha_{avg} - \alpha)dt + \alpha\sigma dz \qquad (9.39)$$

In this equation it is expected that the α value reverts to α_{avg}, but the variance rate grows with α. Here, η is the speed of reversion, and α_{avg} is the "normal" level of α, that is, the level which tends to revert. In order to predict the constants in Equation (9.39), a regression analysis can be performed using the available discrete time data similar to the ideal system presented earlier. We can write this equation in the discrete form as follows.

$$\alpha_t = \eta\,\alpha_{avg}\,\Delta t + (1 - \eta\,\Delta t)\,\alpha_{t-1} + \sigma\,\alpha_{t-1}\,\epsilon_t\,\sqrt{\Delta t} \qquad (9.40)$$

If we compare the equations for geometric Brownian motion (Equation (9.38)) and geometric mean reverting process (Equation (9.40)), we can see that these equations differ from each other by the constant term $\eta\,\alpha_{avg}\,\Delta t$. This constant term reflects the reversion trend. Using this equation, the sample paths for the mean reverting process for a different set of random numbers (ϵ_t in Equation (9.40)) were drawn from a unit normal distribution, as shown in Figure 9.13b. Figures 9.13a and 9.13b confirm that the relative volatility of this mixture can be represented by the geometric mean reverting process.

9.4 Optimal Control under Uncertainty

There are two types of problems in batch distillation which involve optimal control under uncertainty. The first is the stochastic optimal control expected value formulation presented by Diwekar and co-workers (Rico-Ramirez et al., 2003; Ulas and Diwekar, 2004, 2006; Ulas et al., 2005) and second one is worst-case robust optimization formulation of Diehl et al. (2006, 2008). These formulations are presented below.

9.4.1 Stochastic Optimal Control

The stochastic Ito processes described earlier converts the deterministic optimal control problem to a stochastic optimal control problem where the objective function is expectation of the functional. Traditional optimal control techniques like calculus of variations, dynamic programming, or maximum principle are not suitable to handle these problems. To solve this problem one has to use Ito's lemma and a stochastic dynamic programming formulation. Recently, Rico-Ramirez and Diwekar (2004) presented a stochastic version of the maximum principle derived from the stochastic dynamic programming formulation. These two formulations are presented below.

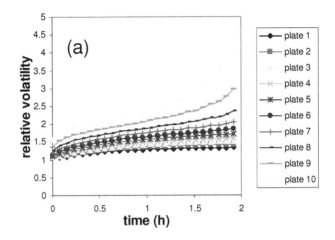

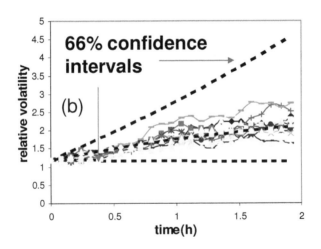

FIGURE 9.13
Relative Volatility as an Ito Process: (a)Relative Volatility Changes for a
Nonideal Mixture, (b) Ito Process Representation (Ulas and Diwekar, 2004)

Ito's Lemma

Ito's lemma is easier to understand as a Taylor series expansion. Suppose that $x(t)$ follows the process of Equation (9.19), and consider a function F that is at least twice differentiable in x and once in t. We would like to find the total differential of this function dF. The usual rules of calculus define this differential in terms of first-order changes in x and t:

$$dF = \frac{\partial F}{\partial t} dt + \frac{\partial F}{\partial x} dx \qquad (9.41)$$

But suppose that we also include higher-order terms for changes in x:

$$dF = \frac{\partial F}{\partial t} dt + \frac{\partial F}{\partial x} dx + \frac{1}{2} \frac{\partial^2 F}{\partial x^2} (dx)^2 + \frac{1}{6} \frac{\partial^3 F}{\partial x^3} (dx)^3 + \ldots \quad (9.42)$$

In ordinary calculus, these higher-order terms all vanish in the limit. For an Ito process following Equation (9.19), it can be shown that the differential dF is given in terms of first-order changes in t and second-order changes in x. Hence, Ito's lemma gives the differential dF as

$$dF = \frac{\partial F}{\partial t} dt + \frac{\partial F}{\partial x} dx + \frac{1}{2} \frac{\partial^2 F}{\partial x^2} (dx)^2 \qquad (9.43)$$

By substituting Equation (9.19) and $(dz)^2 = dt$ in Equation (9.43) and neglecting terms containing $(dt)^2$ and $dtdz$, an equivalent expression is obtained:

$$dF = \left[\frac{\partial F}{\partial t} + a(x,t)\frac{\partial F}{\partial x} + \frac{1}{2}b^2(x,t)\frac{\partial^2 F}{\partial x^2} \right] dt + b(x,t)\frac{\partial F}{\partial x} dz \qquad (9.44)$$

Compared to the chain rule for differentiation in ordinary calculus (Equation (9.41)), Equation (9.43) has one extra term that captures the effect of convexity or concavity of F.

Dynamic Programming Formulation

We have seen that for the deterministic case when no uncertainty is present, the principle of optimality states that the minimum value is a function of the initial state and the initial time, resulting in the Hamilton-Jacobi-Bellman equation. The H-J-B equation states that, for the optimal control problem:

$$\text{Maximize} \quad J = j(\bar{x}_t) + \int_0^T k(\bar{x}_t, \theta_t) \, dt \qquad (9.45)$$
$$\theta_t$$

subject to

$$\frac{d\bar{x}_t}{dt} = f(\bar{x}_t, \theta_t) \qquad (9.46)$$

The optimality conditions are given by:

$$0 = \frac{\partial J}{\partial t} + \text{Maximize}_{\theta_t} \left[k(\bar{x}_t, \theta_t) + \sum_i \frac{\partial J}{\partial x_i} \frac{dx_i}{dt} \right] \qquad (9.47)$$

$$0 = \frac{\partial J}{\partial t} + \text{Maximize}_{\theta_t} \left[k(\bar{x}_t, \theta_t) + \sum_i \frac{\partial J}{\partial x_i} f(\bar{x}_t, \theta_t) \right] \qquad (9.48)$$

where i represents the state variables in the problem.

On the other hand, when uncertainty is present in the calculation, the H-J-B equations are modified to obtain the following objective function.

$$\text{Maximize}_{\theta_t} \quad J = E \left[j(\bar{x}_t) + \int_0^T k(\bar{x}_t, \theta_t) \, dt \right]$$

where E is the expectation operator. If the state variable i can be represented as an Ito process given by Equation (9.49) then the optimality conditions are given by Equation (9.50).

$$dx_i = f_i(\bar{x}_t, \theta_t) \, dt + \sigma_i \, dz \qquad (9.49)$$

$$0 = \text{Maximize}_{\theta_t} \left[k(\bar{x}_t, \theta_t) + \frac{1}{dt} E(dJ) \right] \qquad (9.50)$$

Following Ito's lemma Equation (9.50) results in:

$$0 = k(\bar{x}_t, \theta_t^*) + \frac{\partial J}{\partial t} + \sum_i \frac{\partial J}{\partial x_i} f_i(\bar{x}_t, \theta_t^*)$$

$$+ \sum_i \frac{\sigma_i^2}{2} \frac{\partial^2 J}{(\partial x_i)^2} + \sum_{i \neq j} \sigma_i \sigma_j \frac{\partial^2 J}{\partial x_i \, \partial x_j} \qquad (9.51)$$

where θ^* represents the optimal solution to the maximization problem.

In Equation (9.49), σ_i is the variance parameter of the state variable x_i.

Note that this definition implicitly restricts our analysis for the cases in which the behavior of the state variables can be represented as an Ito process. Also, the extra terms in Equation (9.51) come from the fact that second-order contributions of stochastic state variables are not negligible (see Equation (9.43) and Ito's lemma). If the uncertain variables are independent of each other then the Equation 9.51 reduces to the following equation.

$$
\begin{aligned}
0 \; = \; & k\left(\bar{x}_t, \theta_t^*\right) + \frac{\partial J}{\partial t} + \sum_i \frac{\partial J}{\partial x_i} f_i\left(\bar{x}_t, \theta_t^*\right) \\
& + \sum_i \frac{\sigma_i^2}{2} \frac{\partial^2 J}{\left(\partial x_i\right)^2}
\end{aligned}
\tag{9.52}
$$

As stated earlier, the solution of a stochastic differential equation is not a value for the function, but it is a probability distribution that varies with time. This is a simplified form of stochastic differential equations.

Stochastic Maximum Principle

Although the mathematics of dynamic programming look different from the maximum principle formulation, in most cases they lead to the same results. As a matter of fact, Rico-Ramirez and Diwekar (2004) showed that, starting from the dynamic programming optimality conditions (H-J-B equations), the derivation of the adjoint equations of the maximum principle can be achieved. This is not surprising and has been reported earlier in Chapter 8. They used the mathematical equivalence between dynamic programming and the maximum principle to extend the maximum principle to the stochastic case. The main aspect of the derivations consists in obtaining expressions for *adjoint equations*. The adjoint equations provide the dynamics of the *adjoint variables* in the maximum principle. For the deterministic case, it is shown that the adjoint variables in the maximum principle are equivalent to the derivatives of the objective function with respect to the state variables of the dynamic programming approach. Such an equivalence was shown for the stochastic case, resulting in the following formulation for the stochastic maximum principle.

$$
\underset{\theta_t}{\text{Maximize}} \quad J \; = \; E\left[j(\bar{x}_t) + \int_0^T k\left(\bar{x}_t, \theta_t\right) \, dt \right]
$$

subject to

$$
d\bar{x}_t \; = \; f(\bar{x}_t, \theta_t, x_s) dt + g dz
\tag{9.53}
$$

$$
\bar{x}_0 \; = \; \bar{x}_{initial}
$$

Application of the maximum principle to the above problem involves the addition of nx adjoint variables z_t (one adjoint variable per state variable), nx adjoint equations, and a Hamiltonian, which satisfies the following relations:

$$H(\bar{z}_t, \bar{x}_t, \theta_t) = \bar{z}_t^T f(\bar{x}_t, \theta_t, x_s) = \sum_{i=1}^{nx} z^{(i)} f_i(\bar{x}_t, \theta_t) + \sum_{i=1}^{nx} \frac{1}{2} g_i^2 \omega_i \quad (9.54)$$

$$\frac{dz^{(i)}}{dt} = -\sum_{j=1}^{n} z^{(j)} \frac{\partial f_j}{\partial x^{(i)}} - \frac{1}{2} \left(g_i^2 \right) \omega_i \quad (9.55)$$

$$\bar{z}_T = \bar{c} \quad (9.56)$$

However, in this formulation additional variables and equations due to uncertainties are added, as given below.

$$\frac{d\omega_i}{dt} = -2\omega_i \frac{\partial f_i}{\partial x^{(i)}} - z^{(i)} \frac{\partial^2 f_i}{\partial x^{(i)}}^2 - \omega_i \frac{1}{2} \left(\frac{\partial^2 (g_i)^2}{\partial x^{(i)}} \right)^2 \quad \omega_i(T) = 0$$

9.4.2   Problems in Batch Distillation

The stochastic optimal control problem presented in the batch distillation literature involves the maximum distillate problem presented in Chapter 8. For this case, the stochastic optimal control problem is expressed as:

$$\text{Maximize} \quad L = E\left[\int_0^T \frac{V}{R_{tU} + 1} \left[1 - \lambda(x_D^* - x_D^{(1)}) \right] dt \right]$$
$$R_{tU}$$

subject to

$$dx^{(1)} = \frac{-V}{R_{tU} + 1} dt + x^{(1)} \sigma_1 dz, \ x^{(1)}(0) = B_0 = F \quad (9.57)$$

$$dx^{(2)} = \frac{V}{R_{tU} + 1} \frac{(x^{(2)} - x_D^{(1)})}{x^{(1)}} dt + x^{(2)} \sigma_2 dz, \ x^{(2)}(0) = x_F^{(1)} \quad (9.58)$$

and the optimality conditions can be stated as

$$\frac{\partial L}{\partial t} + \text{Maximize} \ \left[k\left(\bar{x}_t, R_{tU} \right) + \frac{1}{dt} E(dL) = 0 \right] \quad (9.59)$$
$$R_{tU}$$

$$\frac{\partial L}{\partial t} + \underset{R_{tU}}{\text{Maximize}} \quad [k\,(\bar{x}_t, R_{tU}) + \frac{\partial L}{\partial x^{(1)}} f_1\,(\bar{x}_t, R_{tU}) + \frac{\partial L}{\partial x^{(2)}} f_2\,(\bar{x}_t, R_{tU})$$

$$+ \frac{\sigma_1^2}{2}\left(x^{(1)}\right)^2 \frac{\partial^2 L}{\left(\partial x^{(1)}\right)^2} + \frac{\sigma_2^2}{2}\left(x^{(2)}\right)^2 \frac{\partial^2 L}{\left(\partial x^{(2)}\right)^2}$$

$$+ \frac{\sigma_1 \sigma_2}{2} x^{(1)} x^{(2)} \frac{\partial^2 L}{\partial x^{(1)}\,\partial x^{(2)}}] \tag{9.60}$$

Note that if we consider that the uncertainty terms in Equations (9.57) and (9.58) are not correlated, the last term can be eliminated. Hence, by substituting Equations (9.57)-(9.58) into Equation (9.60) we get

$$0 = \frac{\partial L}{\partial t} + \underset{R_{tU}}{\text{Max.}} \quad \frac{V}{R_{tU} + 1}\left[1 - \lambda(x_D^* - x_D^{(1)})\right] - \frac{V}{R_{tU} + 1}\frac{\partial L}{\partial x^{(1)}} +$$

$$\frac{V(x^{(2)} - x_D^{(1)})}{(R_{tU} + 1)x^{(1)}}\frac{\partial L}{\partial x^{(2)}} + \frac{\sigma_1^2}{2}\left(x^{(1)}\right)^2 \frac{\partial^2 L}{\left(\partial x^{(1)}\right)^2} + \frac{\sigma_2^2}{2}\left(x^{(2)}\right)^2 \frac{\partial^2 L}{\left(\partial x^{(2)}\right)^2}$$

$$0 = \left[1 - \lambda(x_D^* - x_D^{(1)}) - \frac{\partial L}{\partial x^{(1)}} + \frac{x^{(2)} - x_D^{(1)}}{x^{(1)}}\frac{\partial L}{\partial x^{(2)}}\right]\left[-\frac{V}{(R_{tU} + 1)^2}\right]$$

$$+ \left[\lambda\frac{\partial x_D^{(1)}}{\partial R_{tU}} - \frac{\partial x_D^{(1)}}{\partial R_{tU}}\frac{1}{x^{(1)}}\frac{\partial L}{\partial x^{(2)}}\right]\left[\frac{V}{R_{tU} + 1}\right]$$

$$+ \sigma_1\frac{\partial \sigma_1}{\partial R_{tU}}\left(x^{(1)}\right)^2 \frac{\partial^2 L}{\left(\partial x^{(1)}\right)^2} + \sigma_2\frac{\partial \sigma_2}{\partial R_{tU}}\left(x^{(2)}\right)^2 \frac{\partial^2 L}{\left(\partial x^{(2)}\right)^2} \tag{9.61}$$

Simplifying, we get an implicit equation for R_{tU}:

$$R_{tU} = \frac{\frac{\partial L}{\partial x^{(2)}}\frac{x^{(2)} - x_D^{(1)}}{x^{(1)}} - \frac{\partial L}{\partial x^{(1)}} - \lambda(x_D^* - x_D^{(1)}) + 1}{\frac{\partial x_D^{(1)}}{\partial R_{tU}}\left[\lambda - \frac{1}{x^{(1)}}\frac{\partial L}{\partial x^{(2)}}\right]}$$

$$- \frac{\left[\sigma_1\frac{\partial \sigma_1}{\partial R_{tU}}\left(x_1\right)^2 \frac{\partial^2 L}{\left(\partial x^{(1)}\right)^2} + \sigma_2\frac{\partial \sigma_2}{\partial R_{tU}}\left(x^{(2)}\right)^2 \frac{\partial^2 L}{\left(\partial x^{(2)}\right)^2}\right]\left[\frac{(R_{tU}+1)^2}{V}\right]}{\frac{\partial x_D^{(1)}}{\partial R_{tU}}\left[\lambda - \frac{1}{x^{(1)}}\frac{\partial L}{\partial x^{(2)}}\right]} - 1$$

Note that if we assume $\sigma_1 = 0$ (i.e., uncertainty exists only in $x^{(2)}$ ($x_B^{(1)}$), but does not exist in $x^{(1)}$ (B)), the equation reduces to

$$R_{tU} = \dfrac{\dfrac{\partial L}{\partial x^{(2)}} \dfrac{x^{(2)} - x_D^{(1)}}{x^{(1)}} - \dfrac{\partial L}{\partial x^{(1)}} - \lambda(x_D^* - x_D^{(1)}) + 1}{\dfrac{\partial x_D^{(1)}}{\partial R_{tU}} \left[\lambda - \dfrac{1}{x^{(1)}} \dfrac{\partial L}{\partial x^{(2)}} \right]}$$
$$- \dfrac{\left[\sigma_2 \dfrac{\partial \sigma_2}{\partial R_{tU}} \left(x^{(2)} \right)^2 \dfrac{\partial^2 L}{\left(\partial x^{(2)} \right)^2} \right] \left[\dfrac{(R_{tU}+1)^2}{V} \right]}{\dfrac{\partial x_D^{(1)}}{\partial R_{tU}} \left[\lambda - \dfrac{1}{x^{(1)}} \dfrac{\partial L}{\partial x^{(2)}} \right]} - 1$$

$$(9.62)$$

Let us think of what we have accomplished for the uncertain case. By assuming that the state variables of the maximum distillate problem can be represented by Equations (9.57) and (9.58), we have obtained an implicit equation (Equation (9.62)) which allows the calculation of the optimal profile for the reflux ratio. However, we had explained before that this work focused on optimal control problems in which the uncertainty in the calculation is introduced by representing the behavior of the relative volatility as a geometric Brownian motion. If so, then why are we assuming that the state variables are the ones that present such an uncertain behavior? We answer this question in the following section. By using Ito's lemma, we show that the uncertainty in the calculation of the relative volatility affects the calculation of one of the state variables ($x^{(2)}$, which is the same as $x_B^{(1)}$), which can also be represented as an Ito process.

State Variable and Relative Volatility: The Two Ito Processes

Recall that, in the quasi-steady-state method of batch distillation optimal control problems considered in this work, the integration of the state variables leads to the calculation of the rest of the variables assumed to be in quasi-steady-state. Also, recall that such variables in quasi-steady-state are determined by applying short cut method calculations.

Let us focus now on the expression for the dynamic behavior of the bottom composition of the key component, Equation (9.58):

$$dx^{(2)} = \dfrac{V}{R_{tU} + 1} \dfrac{(x^{(2)} - x_D^{(1)})}{x^{(1)}} \, dt + x^{(2)} \sigma_2 \, dz \qquad (9.63)$$

The question here is how to calculate the term corresponding to uncertainty in α. To relate the relative volatility to the state variable $x^{(2)}$ ($x_B^{(1)}$), we have to consider the HG equation, which relates the relative volatility to the bottom composition $x_B^{(1)}$ through the constant C_1:

$$1 = \sum_{i=1}^{n} \left(\dfrac{\alpha_i}{\alpha_1} \right)^{C_1} \dfrac{x_D^{(1)}}{x_B^{(1)}} x_B^{(i)} \qquad (9.64)$$

Note that the equation contains the relative volatility to the power of C_1. Rearranging,

$$1 = \frac{x_D^{(1)}}{x_B^{(1)}} \alpha_1^{-C_1} \sum_{i=1}^{n} \alpha_i^{C_1} x_B^{(i)} \tag{9.65}$$

Taking the derivatives of this expression implicitly with respect to $x_B^{(i)}$ and $\alpha_i^{C_1}$,

$$x_B^{(i)} d\alpha_i^{C_1} + dx_B^{(i)} \alpha_i^{C_1} = 0$$

$$\frac{d\alpha_i^{C_1}}{\alpha_i^{C_1}} = -\frac{dx_B^{(i)}}{x_B^{(i)}} \tag{9.66}$$

If we express the behavior of relative volatility by the general equation for an Ito process:

$$d\alpha = f_1(\alpha, t)dt + f_2(\alpha, t)dz \tag{9.67}$$

For the geometric Brownian motion and the geometric mean reverting process, $f_2(\alpha, t)$ is the same for both of these processes. Therefore we can write Equation (9.67) in the following form.

$$d\alpha = f_1(\alpha, t)dt + \sigma\alpha dz \tag{9.68}$$

Then, by using Ito's lemma (Equation (9.43)):

$$dF = \frac{\partial F}{\partial t} dt + \frac{\partial F}{\partial x} dx + \frac{1}{2} \frac{\partial^2 F}{\partial x^2} (\sigma)^2(x)^2 dt$$

we can obtain an expression for the relative volatility to the power of C_1, α^{C_1},

$$d\alpha^{C_1} = \frac{\partial \alpha^{C_1}}{\partial \alpha} d\alpha + \frac{1}{2}\sigma^2\alpha^2 \frac{\partial^2 \alpha^{C_1}}{\partial \alpha^2} dt \tag{9.69}$$

Simplifying:

$$d\alpha^{C_1} = C_1\alpha^{C_1-1}d\alpha + \frac{1}{2}\sigma^2\alpha^{C_1}C_1(C_1-1)dt$$

$$\frac{d\alpha^{C_1}}{\alpha^{C_1}} = C_1\left[\frac{f_1(\alpha,t)}{\alpha}dt + \sigma dz\right]dt + \frac{1}{2}\sigma^2 C_1(C_1-1)dt \tag{9.70}$$

$$= f_{new}(\alpha,t)dt + \sigma_{new}dz \tag{9.71}$$

where

$$f_{new}(\alpha, t) = C_1 \frac{f_1(\alpha, t)}{\alpha} + \frac{1}{2} \sigma^2 C_1 (C_1 - 1)$$

and

$$\sigma_{new} = C_1 \sigma$$

Substituting Equation (9.71) in Equation (9.66) implies that

$$\frac{dx^{(2)}}{x^{(2)}} = \frac{dx_B^{(1)}}{x_B^{(1)}} = -f_{new}(\alpha, t)dt + \sigma_{new}dz \qquad (9.72)$$

Note that Equation (9.72) establishes that the uncertain behavior for the relative volatility results in a similar behavior for the dynamics of $x^{(2)}$. That is, if α is an Ito process, then $x^{(2)}$ is represented by a similar Ito process. For an ideal system, this process is shown to be a geometric Brownian motion, whereas for a nonideal system such as ethanol-water it is found to be a geometric mean reverting process.

Coupled Maximum Principle and the NLP Approach for the Uncertain Case

Although Ito's lemma and dynamic programming helped us to provide an analytical expression for the reflux ratio profile, these equations are cumbersome and computationally inefficient to solve. One of the fastest and simplest methods to solve optimal control problems in batch distillation with no uncertainty is the coupled maximum principle and NLP approach described earlier. Such an approach can also be used in this work for the solution of the optimal control problem in the uncertain case, but in order to do that, the derivation of the appropriate adjoint equations is required. In this section, we show the maximum principle formulation that results from the analysis of the uncertain case (similar to the formulation presented earlier for the deterministic case; we are not considering the Lagrangian expression of the objective function). For details of the stochastic maximum principle general case, please refer to Rico-Ramirez and Diwekar (2004).

The problem is expressed as

$$\underset{R_{tU}}{\text{Maximize}} \quad -x^{(1)}(T) \qquad (9.73)$$

subject to

$$\frac{dx^{(1)}}{dt} = \frac{-V}{R_t + 1}, \quad x^{(1)}(0) = B_0 = F \qquad (9.74)$$

$$dx^{(2)} = \frac{V}{R_{tU} + 1} \frac{(x^{(2)} - x_D^{(1)})}{x^{(1)}} \, dt + x^{(2)} \, \sigma_2 \, dz, \quad x^{(2)}(0) = x_F^{(1)} \quad (9.75)$$

The Hamiltonian, which should be maximized, is:

$$H = \frac{V}{R_{tU} + 1} \frac{\partial L}{\partial x^{(1)}} + \frac{V}{R_{tU} + 1} \frac{\left(x^{(2)} - x_D^{(1)} \right)}{x^{(1)}} \frac{\partial L}{\partial x^{(2}} + \frac{\sigma_2^2}{2} \left(x^{(2)} \right)^2 \frac{\partial^2 L}{\left(\partial x^{(2)} \right)^2}$$
$$(9.76)$$

The adjoint equations are:

$$\frac{dz^{(1)}}{dt} = z^{(2)} \frac{V \left(x^{(2)} - x_D^{(1)} \right)}{(R_{tU} + 1)(x^{(1)})^2}, \quad z^{(1)}(T) = -1 \quad (9.77)$$

$$\frac{dz^{(2)}}{dt} = -z^{(2)} \frac{V \left(1 - \frac{\partial x_D^{(1)}}{\partial x^{(2)}} \right)}{(R_{tU} + 1)x^{(1)}} - \sigma_2^2 x^{(2)2} \frac{\partial^2 L}{\left(\partial x^{(2)} \right)^2}, \quad z^{(2)}(T) = 0 \quad (9.78)$$

Recall that:

$$\frac{\partial L}{\partial x^{(1)}} = z^{(1)}$$

$$\frac{\partial L}{\partial x^{(2)}} = z^{(2)}$$

Also, if we define

$$\frac{\partial^2 L}{\left(\partial x^{(2)} \right)^2} = \omega_t$$

then it can be shown that:

$$\frac{d\omega_t}{dt} = -\omega_t \frac{V \left(1 - \frac{\partial x_D^{(1)}}{\partial x_2} \right)}{(R_{tU} + 1)x^{(1)}} + z^{(2)} \frac{V \left(1 - \frac{\partial^2 x_D^{(1)}}{\left(\partial x^{(2)} \right)^2} \right)}{(R_{tU} + 1)x^{(1)}}$$
$$-\omega_t \, \sigma_2^2 - 2\sigma_2^2 x^{(2)} \frac{\partial^3 L}{\left(\partial x^{(2)} \right)^3}, \quad \omega_T = 0$$

$$(9.79)$$

The optimality conditions on the reflux ratio result in:

$$R_{tU} = \frac{-\frac{\partial L}{\partial x^{(2)}}\frac{x^{(2)}-x_D^{(1)}}{x^{(1)}}+\frac{\partial L}{\partial x^{(1)}}}{\frac{\partial x_D^{(1)}}{\partial R_{tU}}\frac{1}{x^{(1)}}\frac{\partial L}{\partial x^{(2)}}} + \frac{\left[\sigma_2\frac{\partial \sigma_2}{\partial R_{tU}}\left(x^{(2)}\right)^2\frac{\partial^2 L}{\left(\partial x^{(2)}\right)^2}\right]\left[\frac{(R_{tU}+1)^2}{V}\right]}{\frac{\partial x_D^{(1)}}{\partial R_{tU}}\frac{1}{x^{(1)}}\frac{\partial L}{\partial x^{(2)}}} - 1 \tag{9.80}$$

Now, if we define

$$\xi = \frac{\frac{\partial^2 L}{\left(\partial x^{(2)}\right)^2}}{\frac{\partial L}{\partial x^{(1)}}} = \frac{\omega_t}{z^{(1)}} \tag{9.81}$$

$$Z = \frac{\frac{\partial L}{\partial x^{(2)}}}{\frac{\partial L}{\partial x^{(1)}}} = \frac{z^{(2)}}{z^{(1)}} \tag{9.82}$$

and consider negligible third partial derivatives, then, without loss of information, Equations (9.77), (9.78), (9.79), and (9.80) can be reformulated as

$$\frac{dZ}{dt} = -Z\frac{V\left(x^{(2)}-x_D^{(1)}\right)}{(R_{tU}+1)(x^{(1)})^2} - Z\frac{V\left(1-\frac{\partial x_D^{(1)}}{\partial x^{(2)}}\right)}{(R_{tU}+1)x^{(1)}} - \sigma_2^2 x^{(2)^2}\xi, \; Z(T)=0 \tag{9.83}$$

$$\begin{aligned}\frac{d\xi}{dt} &= -\xi\frac{V\left(1-\frac{\partial x_D^{(1)}}{\partial x^{(2)}}\right)}{(R_{tU}+1)x^{(1)}} + Z\frac{V\frac{\partial^2 x_D^{(1)}}{\left(\partial x^{(2)}\right)^2}}{(R_{tU}+1)x^{(1)}} \\ &\quad -\sigma_2^2\,\xi - \xi Z\frac{V\left(x^{(2)}-x_D^{(1)}\right)}{(R_{tU}+1)(x^{(1)})^2}, \; \xi_T = 0\end{aligned} \tag{9.84}$$

$$R_{tU} = \frac{x^{(1)}-Z(x^{(2)}-x_D^{(1)})}{\frac{\partial x_D^{(1)}}{\partial R_{tU}}z} + \frac{x^{(1)}\left[\sigma_2\frac{\partial \sigma_2}{\partial R_{tU}}\left(x^{(2)}\right)^2\xi\right]\left[\frac{(R_{tU}+1)^2}{V}\right]}{\frac{\partial x_D^{(1)}}{\partial R_{tU}}Z} - 1 \tag{9.85}$$

This representation allowed us to use the coupled maximum principle-NLP solution algorithm. In such an approach, the Lagrangian formulation of the objective function is not used in the solution. Most important of all, the algorithm avoids the solution of the two-point boundary value problem for the pure maximum principle formulation, or the solution of partial differential equations for the pure dynamic programming formulation. Note that Equation (9.85) is obtained by maximizing the Hamiltonian (maximizing the distillate), and does not incorporate the purity constraint. Hence, the use of the final boundary condition ($Z_T = 0$, $\xi_T = 0$) provides the limiting solution resulting in all the reboiler charge instantaneously going to the distillate pot ($R = -\infty$) with the lowest overall purity. Because in this approach the purity constraint is imposed externally to the Hamiltonian, then the final boundary condition is no longer valid. Instead, the final boundary condition is automatically imposed when the purity constraint is satisfied. The algorithm involves the solution of the NLP optimization problem for the scalar variable R_0, the initial reflux ratio, subject to:

1. The dynamics of the state variables given by Equations (9.74) and (9.75)

2. The adjoint equations (Equations (9.83) and (9.84)) and the initial conditions for these adjoint equations, derived in terms of the decision variable R_0.

3. The optimality conditions for the control variable (reflux ratio, Equation (9.85)).

Earlier, it was established by numerical experiments that the uncertainties in relative volatility can be represented as an Ito process. For the optimal control problem, the system considered is 100 kmol of ethanol-water from Ulas and Diwekar (2004) being processed in a batch column with 1 atm pressure, 13 theoretical stages, 33 kmol/h vapor rate, and a batch time of 2 hours. For this problem, the purity constraint on the distillate is specified as 90%. The optimal reflux profiles and optimal distillate flow rates for the stochastic case and the deterministic case are shown in Figure 9.14. There is a significant difference between the two profiles. These two profiles for the reflux ratio are given to a rigorous simulator ($MultiBatch\underline{DS}^{TM}$, Diwekar, 1996) to compare the process performances. The average purity is found to be almost the same at about 90% for both of these cases. However, for the deterministic case the distillate amount is 69% lower than the stochastic case. This case study shows that representing uncertainties in relative volatility with Ito processes can significantly improve the system performance in terms of product yield.

9.4.3 Worst-Case Robust Optimization Formulation

Diehl et al. (2006) and Diehl et a. (2008) presented the worst-case robust optimization problem for batch distillation. They considered discretized linear reflux ratio profiles to solve the underlying optimal control problem, converting it into an nonlinear programming (NLP) formulation. As stated earlier, the worst-case robust optimization problem involves a min-max formulation. Diehl et al. (2006) presented an approximate formulation based on linearization of objective function and constraints. Diehl et al. (2008) used a different computational strategy that involved solution by tracking of worst-case solutions. Here we present the first approach because of its simplicity. This will provide

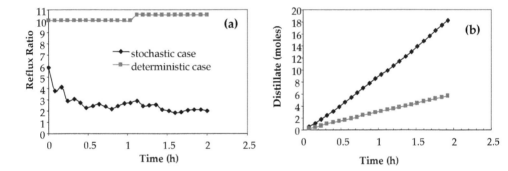

FIGURE 9.14
Optimal Profiles for the Deterministic and Stochastic Cases (Ulas and Diwekar, 2004).

insight into the worst-case robust optimization problem for batch distillation. In this approach Diehl et al. (2006) approximated the bi-level optimization problem given by Equations (9.10-9.15) with a single optimization problem where the inner optimization problem approximation is used. They assumed that the inner optimization problem is convex. With this assumption, they could approximate the problem given by Equations (9.10-9.15) as follows.

$$\text{Min}\quad J \;=\; j(x,u,\theta) + \text{Max}\; \frac{\partial j}{\partial x}(x,u,\theta)\Big(-\big(\frac{\partial h}{\partial x}(x,u,\theta)\big)^{-1}\frac{\partial h}{\partial u}(x,u,\theta)\Big)\triangle u$$

$$x,u \qquad\qquad \theta \qquad s.t. \; \|\triangle u\| \leq 1 \qquad\qquad\qquad (9.86)$$

$$\phi(\theta) \;=\; \text{Max}\frac{\partial g}{\partial x}(x,u,\theta)\Big(-\big(\frac{\partial h}{\partial x}(x,u,\theta)\big)^{-1}\frac{\partial h}{\partial u}(x,u,\theta)\Big)\triangle u$$

$$\qquad\quad \theta \qquad s.t. \; \|\triangle u\| \leq 1 \qquad\qquad\qquad (9.87)$$

$$\phi(\theta) \;\leq\; 0 \qquad\qquad\qquad (9.88)$$

$$h(x,u,\theta) \;=\; 0 \qquad\qquad\qquad (9.89)$$

where θ is the control variable.

For the batch distillation worst-case robust optimization problem, both the approaches used a combination of minimum time and maximum distillate as their objective given by $T - D(T)$. Uncertainties are assumed in the parameter α, which is equivalent to the relative volatility and initial feed composition x_F. The purity is assumed to be 99%. They assumed relative error in α equal to 50% and 10% for x_F. In both approaches for batch distillation, they discretized the problem into 15 equally spaced intervals on the variable time horizon. Figure 9.15 shows the reflux ratio profiles for nominal deterministic optimization and robust optimization. The time required for a nominal solution was found to be equal to $T = 1.3$, while the distillate for this solution was found to be $D(T) = 51$, resulting in the objective function value as -49.7. From Figure 9.16a it can be seen that for the active purity constraint, if the composition is reduced by 0.05, the final distillate composition is reduced by 7.5%. This means that in the linearied worst-case a purity of only 91.5% will be reached. On the other hand, for the robust case profile shown in Figure 9.15b, the sensitivity of purity with respect to uncertain variables shown in Figure 9.16b is reduced considerably. The objective function for this formulation reached the value of -49 ($T = 1.4; D(T) = 50.4$). However, the distillate purity is 99.8% instead of 91.5% reached by the deterministic formulation.

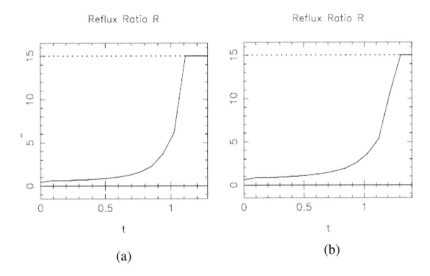

FIGURE 9.15
Optimal Reflux Profiles for (a) Deterministic and (b) Worst-Case Robust Optimization(Diehl et al., 2006)

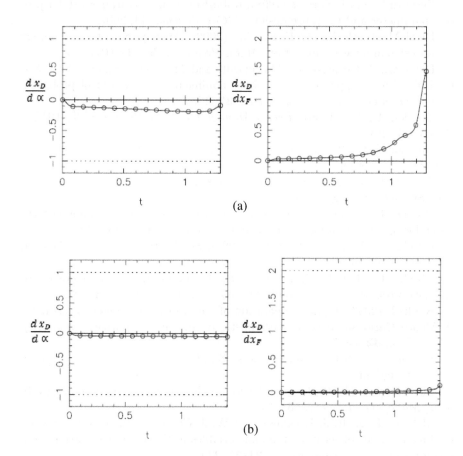

FIGURE 9.16
Sensitivity Profiles for (a)Deterministic and (b) Worst-Case Robust Optimization (Diehl et al., 2006)

References

Arellano-Garcia H., W. Martini, M. Wendt, P. Li, and G. Wozny (2003), Chance constrained batch distillation process optimization under uncertainty, *Proceedings of Foundations of Computer Aided Process Operations 2003*.

Barolo M. and P. Cengio (2001), Closed-loop optimal operation of batch distillation columns, *Computers and Chemical Engineering*, **25**, 561.

Bernard F. and P. Saraiva (1998), Robust optimization framework for process parameter and tolerance design, *AIChE Journal*, **44**, 2007.

Diehl M., H. Bock, and E. Kostina (2006), An approximation technique for robust nonlinear optimization, *Math. Program.*, **Ser. B 107**, 213.

Diehl M., J. Gerhard, W. Marquardt, and M. Monnigmann (2008), Numerical solution approaches for robust nonlinear optimal control problems, *Computers and Chemical Engineering*, **32**, 1287.

Diwekar U. M. (1996), *User's Manual for MultiBatchDS^{TM}*, BPRC, Pittsburgh, PA.

Diwekar U. M. (2003), *Introduction to Applied Optimization*, Kluwer, Dordrechdt, The Netherlands.

Diwekar U. M. (2008), *Introduction to Applied Optimization, Second Edition*, Springer, Springer.com.

Diwekar U. M. and J. R. Kalagnanam (1997), An efficient sampling technique for optimization under uncertainty, *AIChE Journal*, **43**, 440.

Diwekar U. M. and E. S. Rubin (1991), Stochastic modeling of chemical processes, *Computers and Chemical Engineering*, **15**, 105.

Diwekar U. M. and E. S. Rubin (1994), Parameter design methodology for chemical processes using a simulator, *Ind. Eng. Chem. Res.*, **33**, 292.

Kooks I. (2003), Optimal operation of batch processes under uncertainty: A Monte Carlo simulation-deterministic optimization approach, *Ind. Eng. Chem. Res.*, **42**, 6815.

Luckacs E. (1970), *Characteristic Functions*, Charles Griffin, London, United Kingdom.

Madansky A. (1960), Inequalities for stochastic linear programming problems, *Management Science*, **6**, 197.

McKay M. D., R. J. Beckman, and W. J. Conover (1979), A comparison of three methods of selecting random variables in the analysis of output from a computer code, *Technometrics*, **21(2)**, 239.

Metropolis N. and S. Ulam (1949), The Monte Carlo method, *J. Am. Stat. Assoc.*, **44(247)**, 335.

Nemhauser G. L., A. H. G. R. Kan, and M. J. Todd (1989), *Optimization: Handbooks in Operations Research and Management Science*, Vol. 1. North-Holland Press, NY.

Rico-Ramirez V. and U. Diwekar (2004), Stochastic maximum principle for optimal control under uncertainty, *Computers & Chemical Engineering*, **28**, 2845.

Rico-Ramirez V., B. Morel, and U. Diwekar (2003), Real option theory from finance to batch distillation, *Computers and Chemical Engineering*, **27**, 1867.

Srinivasan B., D. Bonvin, E. Visser, and S. Palanki (2002), Dynamic optimization of batch processes II. Role of measurements in handling uncertainty, *Computers and Chemical Engineering*, **27**, 27.

Taguchi G. (1986), *Introduction to Quality Engineering*, Asian Productivity Center, Tokyo, Japan.

Tintner G. (1955), Stochastic linear programming with applications to agricultural economics, *Proc. 2nd Symp. Lin. Progr.*, Washington, DC, s197.

Ulas S. and U. Diwekar (2004), Thermodynamic uncertainties in batch processing and optimal control, *Computers and Chemical Engineering*, **28**, 2245.

Ulas S., U. Diwekar, and M. Stadtherr (2005), Uncertainties in parameter estimation and optimal control in batch distillation, *Computers and Chemical Engineering*, **29**, 1805.

Ulas S. and U. Diwekar (2006), Integrating product and process design with optimal control: A case study of solvent recycling, *Chemical Engineering Science*, **61**, 2001.

Vajda S. (1972), *Probabilistic Programming*, Academic Press, New York.

10

BATCH DISTILLATION SOFTWARE PROGRAMS

CONTENTS

In this chapter, I compare three commercial software programs for batch distillation, namely, the batch distillation unit operation module from CHEMCAD 6.3.2, ASPEN Batch Distillation V. 7.1(BATCHSEP), and the MultiBatch*DS* program from the author.

10.1 CHEMCAD Batch Distillation

Figure 10.1 shows the batch distillation module in the CHEMCAD simulator. It uses CHEMCAD databank and thermodynamic packages. The interface is simple to use. Figure 10.2 shows the specification scheme for this module. Initial operation is assumed to be the total reflux condition. However, it is not possible to simulate this mode of operation except the end steady state condition. Figure 10.3 shows the pot input and Figure 10.4 operating conditions for various batch fractions. This is the simplest module which can simulate constant reflux operation of the conventional batch column. It has a very nice interface for plotting transient profiles as shown in Figure 10.5. The module is for simulation of the conventional batch distillation column with single or multiple fractions.

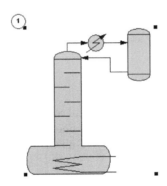

FIGURE 10.1
Batch Distillation Module in CHEMCAD

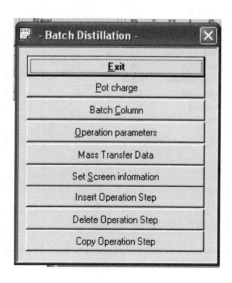

FIGURE 10.2
First Input Screen for the Module

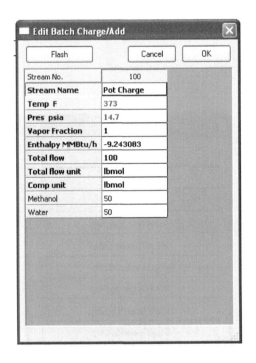

FIGURE 10.3
Pot Input

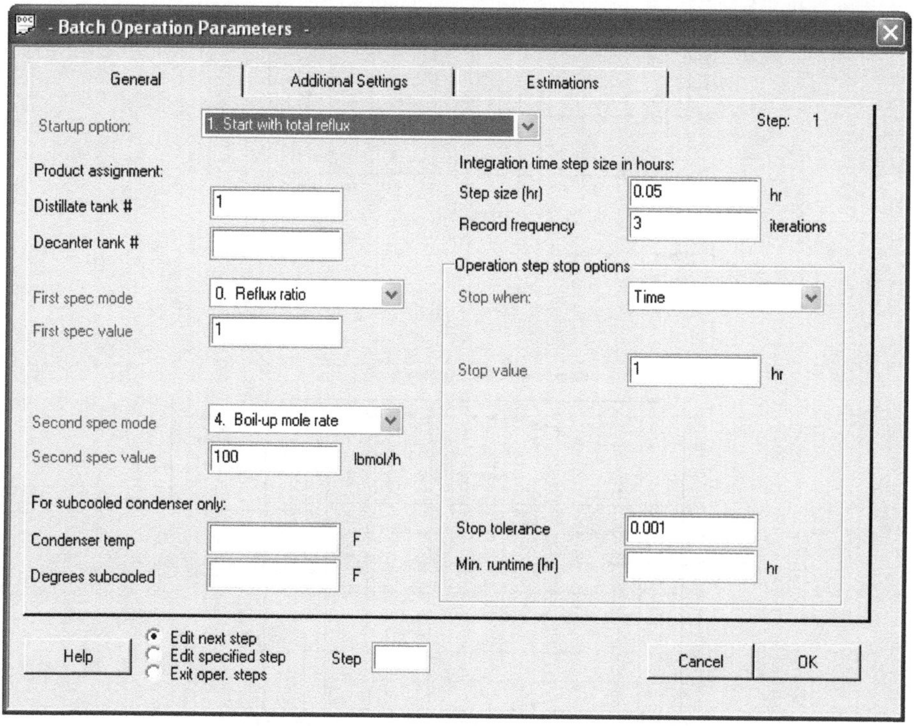

FIGURE 10.4
Operating Conditions Input

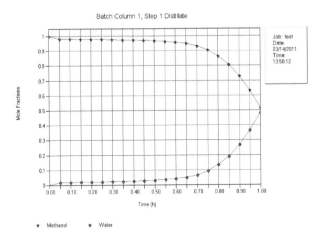

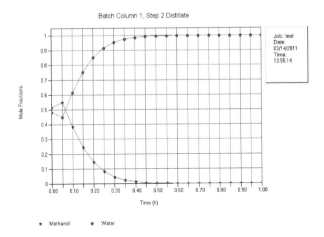

FIGURE 10.5
Transient Profiles

10.2   ASPEN TECH: BATCHSEP

This is a separate package you can buy from ASPEN Technologies Inc. The required input is shown in the left hand side panel of Figure 10.6. The key features of the package are:

- Aspen Batch Distillation uses the same advanced physical property system used by Aspen Plus.

- The ability to define a sequence of operating steps defining how the column is run.

- The ability to model controllers for variable reflux operation, as shown in Figure 10.7.

- The ability to simulate fixed equation optimal reflux policy.

- Charging material at any time to model start-up from empty and fed-batch operation.

- Options for fixed heating rate or prediction of heating rate based upon pot geometry, liquid level, and heating medium conditions, as shown in Figure 10.8.

- Support for distillation with reaction for both 2- and 3-phase systems.

However, the plot interface for this package is not as automatic as the batch distillation module in CHEMCAD or MultiBatch*DS*. Most of the results need to be cut and pasted from the tables generated at the interface, as shown in Figure 10.9. Further, it is not possible to simulate low holdup columns with this package. Similar to the batch distillation module in CHEMCAD, this package cannot simulate equilibration time or the dynamics of the initial total reflux condition, which can be useful in optimization. The package is suitable for simulating an existing column and is not very useful for designing a new column where data needed for this package is not available. Again, this package is restricted to conventional batch column operation with a single or multiple fraction as well as for simulating a recycled waste cut operation.

10.3   BPRC: MultiBatch*DS*

This is the latest package for batch distillation; therefore, it has the ability to simulate and design various configurations of batch distillation like the stripper and the middle vessel column along with the conventional batch distillation,

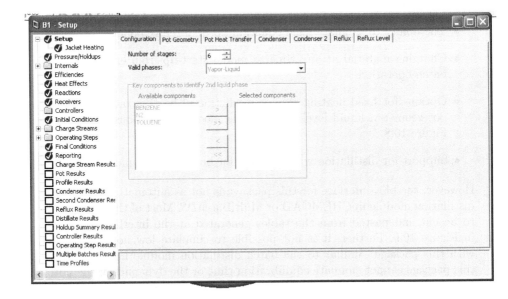

FIGURE 10.6
Required Input for BATCHSEP

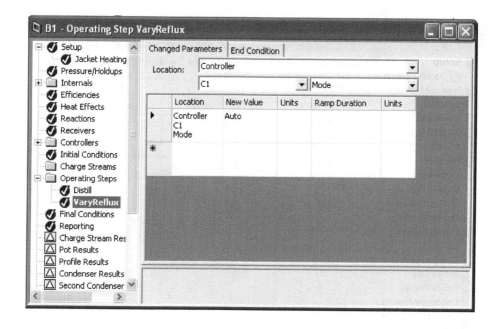

FIGURE 10.7
Variable Reflux Operating Step

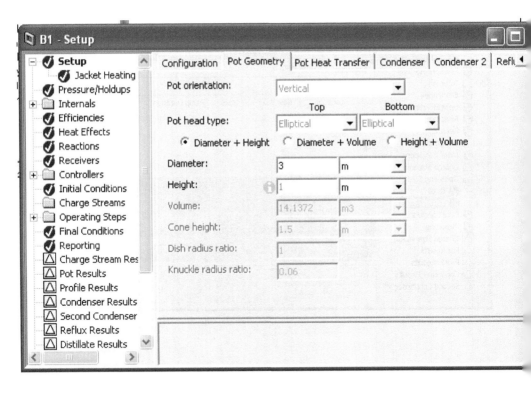

FIGURE 10.8
Pot Geometry and Other Detail Input

2. Click **Plot** as the form type, and in the **Form Name** field, type a name for your plot. Click **OK.**

This creates a new empty time plot.

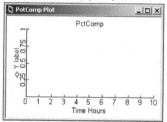

3. Open the **result** form to show the value that you want to plot. Click on this value to select it.

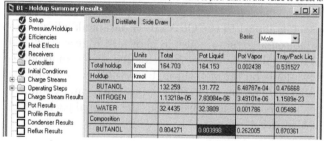

FIGURE 10.9
Plot Interface for BATCHSEP

semi-batch operation, and the recycle waste cut operation. Figure 10.10 shows the input for the three different column configurations. The three operating modes for the conventional batch column, namely, constant reflux, variable reflux, and optimal reflux operating mode simulation (Figure 10.11), and a similar three operating modes for the stripper are possible with this package. These two specifications for the middle vessel column result in 9 operating modes. All these modes can be simulated with this package (Figures 10.10 and 10.11). Multiple fractions, multiple side streams and products (Figure 10.12), and recycle waste cut (Figure 10.13) are possible.

Initial total reflux and total reboil conditions, can be simulated as a steady state or dynamic operation followed by various fractions as shown through the screen shown in Figures 10.12 and 10.13. Figure 10.14 shows transient profiles (left hand side, total reflux dynamics stopping at the equilibration time and right hand side, first fraction dynamics) for these operations.

A hierarchy of models is available with this package starting from the shortcut method for design, optimization, control, and simulation, low holdup semi-rigorous model, reduced order collocation based model, and the rigorous model (Figure 10.15). Heat balance can be turned on and off depending on the availability of data and requirements. Comparison of various models is possible and can be visualized as shown in Figure 10.16. Various column configurations can also be compared. Figure 10.17 shows comparison of the middle vessel distillate composition profiles with that of conventional column profiles and Figure 10.18 shows comparison between the middle vessel column and the stripper.

Optimization and uncertainty analysis (Figures 10.19 and 10.20-10.21) are available with this package.

The package is linked to the CRANIUM databank where data can be obtained from molecular structure also as shown in Figure 10.22 (CRANIUM is a product of Molecular Knowledge Systems Inc. (www.molknow.com). Multi-Batch*DS* is available on the web through www.vri-custom.org or through the author (urmila@vri-custom.org). Both educational and commercial versions are available. This package is derived from the original package BATCH-DIST described in Chapter 4.

10.4   Comparison of Software Packages

Table 10.1 presents the comparison of these three packages briefly.

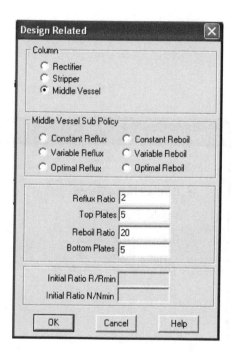

FIGURE 10.10
Column Input for Three Different Configurations

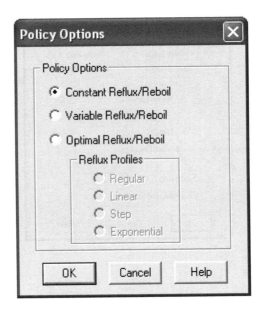

FIGURE 10.11
Policy Options

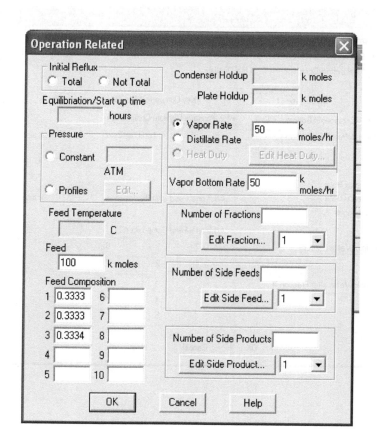

FIGURE 10.12
Multiple Fractions, Side Feeds, and Side Products

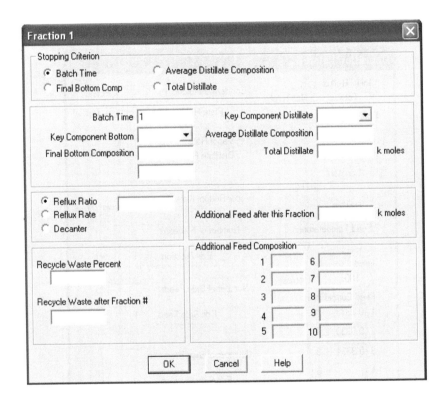

FIGURE 10.13
Recycle Waste Cut and Semi-Continuous Operation

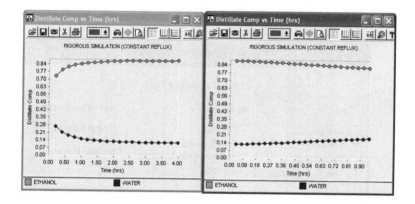

FIGURE 10.14
Transient Profiles for Total Reflux Followed by First Fraction

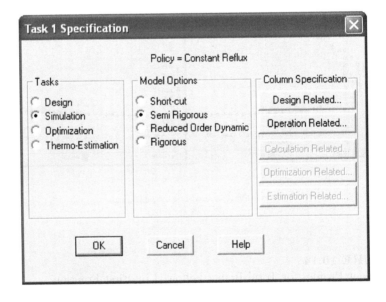

FIGURE 10.15
Various Models in MultiBatch*DS*

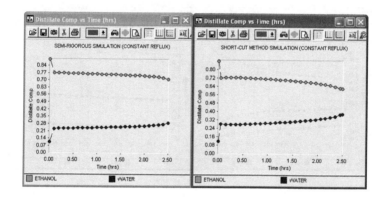

FIGURE 10.16
Comparison of the Shortcut with the Low Holdup Rigorous Model

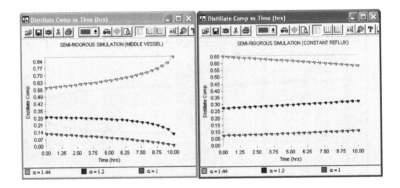

FIGURE 10.17
Comparison of the Middle Vessel Column vs. the Conventional Column

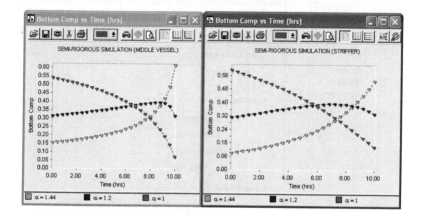

FIGURE 10.18
Comparison of the Middle Vessel Column vs. the Stripper

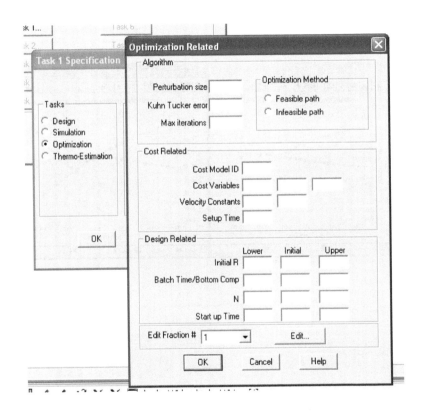

FIGURE 10.19
Optimization Input

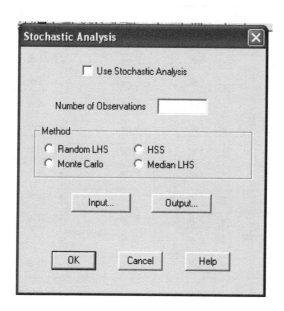

FIGURE 10.20
Uncertainty Analysis Input

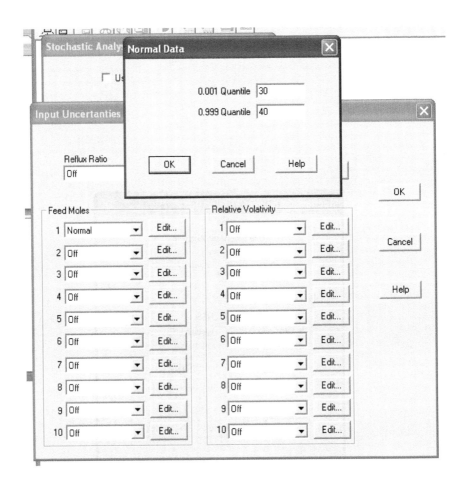

FIGURE 10.21
Uncertainty Analysis Input

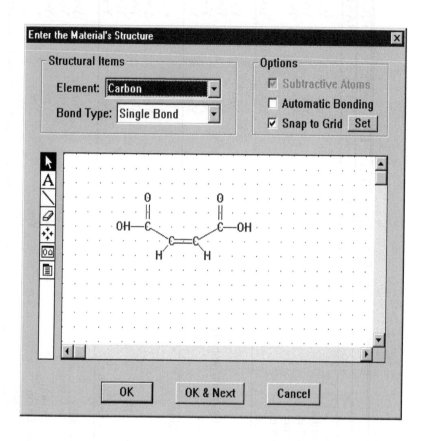

FIGURE 10.22
Structural Input in CRANIUM

TABLE 10.1
Comparison of Software Packages

Features	CHEMCAD BATCH CHEMCAD	BATCHSEP ASPEN PLUS	MultiBatch*DS*	CRANIUM
Databank				
Operations				
Constant Reflux	Yes	Yes	Yes	Yes
Variable Reflux	No	Yes	Yes	Yes
Optimal Reflux	No	No	Yes	Yes
Optimal Reflux				
Fixed Equation	No	Yes	Yes	Yes
Models				
Shortcut	No	No	Yes	Yes
Low Holdup Rigorous/Semirigorous	Yes	No	Yes	Yes
Reduced Order	No	No	Yes	Yes
Rigorous	Yes	Yes	Yes	Yes
Configurations				
Rectifier	Yes	Yes	Yes	Yes
Semi-batch	No	Yes	Yes	Yes
Recycle waste cut	No	Yes	Yes	Yes
Stripper	No	No	Yes	Yes
Middle Vessel Column	No	No	Yes	Yes
Options				
Design Feasibility	No	No	Yes	Yes
Optimization	No	Yes	Yes	Yes
Reactive Distillation	No	Yes	Yes	Yes
3 phase Distillation	Yes	Yes	Yes	Yes
Uncertainty Analysis	No	No	Yes	Yes

Index

A-stable, 65
A-stable systems, 50
Adams–Bashforth
     formula, 48
     method, 47, 48
Adams–Moulton
     formula, 48
     method, 47, 48
Addison, 50, 61
adjoint
     equation, 269, 270, 275, 280, 283, 284, 319, 320, 324, 325, 327
     variable, 269, 271, 277, 280, 283, 284, 319, 320
Aggrawal, 124
Ahmad, 184, 186, 187, 251
Akgiray, 264, 268, 288
Al-Tuwaim, 68, 104
Albrecht, 51, 61
Alexander, 51, 61
Alvarez-Ramirez, 288, 290
Anderson, 251
Arellano-Garcia, 305, 332
Aris, 271, 273, 288
ASPEN, 335, 341, 360
azeotropic, 159
     column, 104
     composition, 173, 236
     continous distillation, 164
     curve, 165
     distillation, 105, 125, 159, 161, 164, 167, 169, 175, 187, 251–253
     mixture, 164, 186, 189–192, 195–197, 201–204, 207, 208, 252
     mixtures, 163

point, 163, 165, 169, 171, 173, 175, 251
reactive distillation, 236
reactive system, 236
system, 11, 90, 121, 163–165, 167, 170, 173–176, 179, 205, 237–240, 253, 306

backward difference formula, BDF, 50
backward Euler's method, 50
Bader, 50, 61
Barolo, 287, 288, 294, 332
Barreto, 212, 251
Barton, 184, 187, 188, 251
batch distillation region, 183–187, 189, 192, 195, 201, 205, 241, 243, 249, 253
BATCH-DIST, 61, 63, 68, 102, 104, 346
BATCHSEP, 335, 341, 342, 345, 360
BDF, 50, 62, 63
Beale, 121, 124
Beckman, 332
Beightler, 108, 124
Bellman, 264, 271, 288, 317
Benallou, 83, 104
Bendda, 252
Benders, 121, 124
Bernard, 305, 332
Bernot, 61, 121, 124, 164, 167, 178, 181, 183, 184, 192, 251
Bernoulli, 107
Berto, 287, 288
Beta distribution, 295
Bickart, 51, 61
Biegler, 125, 258, 264, 268, 289, 290

For Product Safety Concerns and Information please contact our
EU representative GPSR@taylorandfrancis.com Taylor & Francis
Verlag GmbH, Kaufingerstraße 24, 80331 München, Germany